With the French Flying Corps

German trenches.   No man's land.   French trenches.

A VIEW OF THE FRENCH AND GERMAN TRENCHES

With the French Flying Corps

The Experiences of an American Pilot During the First World War

Carroll Dana Winslow

LEONAUR

With the French Flying Corps
The Experiences of an American Pilot During the First World War
by Carroll Dana Winslow

First published under the title
With the French Flying Corps

Leonaur is an imprint of Oakpast Ltd

Copyright in this form © 2011 Oakpast Ltd

ISBN: 978-0-85706-713-5 (hardcover)
ISBN: 978-0-85706-714-2 (softcover)

http://www.leonaur.com

Publisher's Notes

Contents

To
My Father

My Enlistment

In the last two years aviation has become an essential branch of the army organization of every country. Daily hundreds of pilots are flying in Europe, in Africa, in Asia Minor; flying, fighting, and dying in a medium through which, ten years ago, (as at time of first publication), it was considered impossible to travel. But though the air has been mastered, the science of aero-dynamics is still in its infancy, and theory and practice are improved so often that even the best aviators experience difficulty in keeping abreast of the times.

My experience in the French Aviation Service early taught me what a difficult and scientific task it is to pilot an aeroplane. By piloting I mean flying understandingly, skilfully; not merely riding in a machine after a few weeks' training in the hope that a safe landing may be made. In America many aviators holding pilot's licenses are in reality only conductors. Some pilots have received their brevets in the brief period of six weeks. I can only say that I feel sorry for them. My own training in France opened my eyes. It showed me how exhaustive is the method adopted by the belligerents of Europe for making experienced aviators out of raw recruits. Time and experience are the two factors essential in the training of the military pilot. Even in France, where the Aviation Service is constantly working under the forced draught of war conditions, no less than from four to six months are devoted to the training of finished pilots.

Although I have just come from France, the progress of aviation is so rapid that much of my own knowledge may be out of date before I again return to the front. But interest in flying is becoming so general among Americans that the way the aviators of France are trained, and what they are accomplishing, should attract more than passing attention. Surely, what France has done, and is doing, should be an object-lesson to our own government.

Through a special channel only recently open to Americans I enlisted in the French Air Service. As is usual in governmental matters, there were many formalities to be complied with, but in my case a friendly official in the Foreign Office came to the rescue and arranged them for me. After a few days I received the necessary permit to report for duty. Without delay I hurried to the recruiting office, which is located in the Invalides, that wonderfully inspiring monument of martial France. As I entered the *bureau* I met a crowd of men who had been declared unfit for the front, either on account of their health, or because they had been too seriously wounded. But to a man they were anxious to serve "*la patrie*," and were seeking to be re-examined for any service in which physical requirements were not so stringent. For an "*embusque*" (a shirker) is looked upon as pariah in France.

When I had signed a contract to "obey the military laws of France and be governed and punished thereby," I received permission to join the French Air Service. With about thirty other men I marched to the doctor's office, where I was put through the eye, lung, and heart test. I was then ordered to report to the sergeant who had charge of the men who had passed the examination.

Among those accepted I noticed a young man of the working class. He had been particularly nervous while the roll was called. But the moment he heard his own name he seemed overjoyed. Outside, on the sidewalk, his wife was waiting. He dashed out to tell her the news. Instead of bursting into tears, as I had rather expected, she seized his hands and they danced down the street as joyfully as two children. It was typical of the spirit of the French women, willing to sacrifice everything, to help bring victory to their country.

I received my service-order to proceed immediately to Dijon, the headquarters of the Flying Corps. I took the first train and arrived there at about three in the morning. I discovered that the offices did not open until seven, and, as I had nothing to do and was hungry, I sought the military buffet at the railway-station. It was filled with men on leave and others who had been discharged from the hospitals, all waiting to return to the front. Officers and men mingled in a spirit of democracy and "*camaraderie.*" This made a deep impression upon me, for, while discipline in the French army is very strict, there is an entire absence of that snobbishness which the average civilian so often associates with a military organization.

About seven o'clock I made my way to the camp. A sentry challenged me, but after I had proved my identity he sent me to the adjutant, who took my papers and, after reading them, addressed me in perfect English. I was surprised and asked him how he happened to speak English so well. It seems that he had lived in New York for twelve years, but on the outbreak of the war had returned at once to serve. I was then given in charge of a corporal. After this I was put through another "*questionnaire.*" One officer asked for my pedigree; to another I gave the name and address of my nearest relative, to be notified in the event of my death. After this came the "*vestiaire.*" Each "*dépôt,*" or head-quarters, has one of these, where every soldier is completely outfitted by the government.

I received a uniform, two pairs of shoes, two pairs of socks, an overcoat, two suits of underwear, two hats, a knapsack, and a tin cup, bowl, and spoon. The recruit may buy his own outfit if he wishes, but the government offers it to him *gratis* if he is not too particular. I was now a full-fledged French soldier of the second class, second because there was no third. My satisfaction was only exceeded by my embarrassment. I felt very self-conscious in my uniform, but, as a matter of fact, I was less conspicuous in this garb than I was before I gave up my civilian clothing.

The adjutant now gave me three cents, my first three days'

pay as a soldier, and warned me "not to spend it all in one place." Aviators receive extra pay, but I was still only a simple "*poilu*." He then handed me a formal order to study aviation—to be an "*élève pilote*," as they say in France—and also a pass to proceed to Pau.

My time was now my own, so I decided to take a look around the hangars, and before long two "*élève pilotes*" greeted me and inquired whether I was entering the Aviation Corps. When they heard that I was, and that I was an American they told me that they also, and several of their friends to whom they afterward introduced me, had lived for some time in the United States. With all this welcome I became conscious of the understood but inexplicable freemasonry that binds all aviators together. I was greeted everywhere as a comrade and shown everything. I was amazed at the vastness of it all and at the scale of the organization. In one corner of the establishment they were teaching mechanics how to repair motors, in another how to regulate aeroplanes. Beyond were classes for chauffeurs, and countless other courses. There must have been several thousand men, and all of them were merely learning to serve the national heroes, the "*aviateurs*."

In the evening we all went to Dijon together. We dined and went to the theatre. The theatre was full of soldiers, and every little while the provost marshal's guard, composed of gendarmes, would enter and make an arrest. Anyone who does not produce papers explaining his absence from the army is hustled off immediately. There are very few Frenchmen who attempt to dodge their service, but this system of supervision has been found necessary to keep down their number and to discover any German spies who may be about.

After the play I went to the station. The road was clogged with troop-trains carrying reinforcements to the Near-Eastern front. During the four hours I spent in the station twelve trains of British artillery passed by. The *ententé* between the Tommies and the French was very cordial. As the trains came to a stop the men would make a rush for the station buffet, and the French would exchange all sorts of pleasantries with them. Right here I had a lot of fun with the Tommies, for they could not understand how a Frenchman could speak English so fluently.

Then came my train, and I found myself *en route* for Pau. As there were already several American "*élève pilotes*" at the aviation school, I had no difficulty in learning the ropes. It was all very simple. But it was well to know what to expect, especially when it came to the question of discipline, which was very strict until one became a full-fledged aviator. It was just like going back to school and I settled down for the long grind.

First Principles

I have rarely been as much impressed as when I first saw the flying school at Pau. It is situated eight miles beyond the town on the hard meadow-lands granted in the sixteenth century to the villages of Osseau by Henri IV. The grant is still in effect, and the fields now in use are only rented by the government. They make a perfect aviation-ground. Four separate camps and a repair-station lie about a mile from one another and are named—Centre, Blériot, etc. At Centre I saw the low, gray hangars that house the aeroplanes, the tall wireless mast over which the *communiqués* from Paris are daily received, the office-building for the captain and monitors of the school, and the little *café* across the road where everyone goes when off duty. Beyond were the Boche prisoners working on the road, building fences, or cutting wood, under direction of their non-commissioned officers and decrepit old territorials—grim reminders that this flying business is not all play.

It was early morning—the mists were slowly lifting—when the "*élève pilotes*" gathered for their daily work. Mechanics ran the machines out on the field in long lines, and the motors woke to motion with startling roars. One by one the pilots stepped in, and one by one the little biplanes moved swiftly across the field, rose, dipped slightly, rose again, and then mounted higher and higher into the gray sky. In the distance the snowy peaks of the Pyrenees formed an impressive background.

At almost any time during training hours one can see from ten to twenty machines in the air. There are over three hundred

MECHANICS RAN THE MACHINES OUT ON THE FIELD IN LONG LINES

men training. The repair-shops are like a large manufacturing plant. Five hundred mechanics are continually employed there. Among these are little Indo-Chinese, or "*Anamites*," as the French call them, who have come from distant Asia to help France in her struggle for liberty. As French citizens they are mobilized and wear the military uniform, but their tasks are usually of the monotonous, routine variety.

The repair-shops are continually working under pressure, as accidents occur daily.

It is estimated that the average cost to France of training each pilot is five thousand dollars. Most of the accidents, however, are caused by carelessness, stupidity, or overconfidence. The day I joined the school two of the members lost their lives in a curious accident. They were flying at a great height, but thoughtlessly allowed their machines to approach too closely. Before they could change direction there was a crash, and both came tumbling to earth. When two aeroplanes come too near to each other the suction of their propellers pulls them together and they become uncontrollable. That is what happened to these two unfortunate "*élèves*."

The officer in charge of the school explained at length just how this accident happened. We were cautioned against overlooking the fact that the speed of an aeroplane is always spoken of in reference to the body of the air in which the machine is moving. Thus an aeroplane travelling eighty miles an hour with a twenty-mile breeze is travelling at a speed of a hundred miles an hour in reference to the ground. The two machines at the time of the accident were flying east and west, but, while both were travelling at the same speed with reference to the ground, the plane moving in the direction of the wind was making about ninety miles an hour, while the other was covering barely fifty miles at the same time.

The speed at which they were approaching one another was, however, approximately one hundred and forty miles an hour, or more than two miles a minute. What, under ordinary circumstances, would have been a safe distance became a danger zone, and before either pilot realized his mistake it was too late to steer clear.

The scene of the accident lay over a part of the field where Wright's Barn stands. This little red building was the workshop of the Wright brothers when they astonished the world by their first aerial flights. Today that little red barn stands as a monument to American stupidity, for when we allowed the Wrights to go abroad to perfect their ideas instead of aiding them to carry on their work at home, we lost a golden opportunity. Now the United States, which gave to the world the first practical aeroplane, is the least advanced in this all-important science, (as at time of first publication).

Although I came to Pau with a little preliminary experience, and had the "feel" for engines and steering, I was obliged to begin, with the others, at the bottom of the primary class. It was nearly two months before I was allowed to make my first flight. The French idea is that before a pupil commences his apprenticeship as a pilot he must understand thoroughly the machine he is going to handle and know just what he is trying to do in the air. Together with twenty-five other men, who began their studies at about the same time, I was ordered to attend the theoretical courses. When not in the classroom we were stationed on the aviation-field where we could watch the more advanced "*élèves*" fly, thus familiarizing ourselves by observation with all the details of our profession. Classroom work and field-practice go hand in hand.

At first I did not realize how important these courses were, or how strict was the discipline under which we lived. One day, when my thoughts were a little more intent upon an expected weekend at Biarritz than upon what was being explained on the blackboard, the lecturer suddenly asked me a question. I could not answer and forthwith my forty-eight-hour leave was retracted. My Sunday was spoiled, but I considered myself lucky not to have received a "*consigne*," which involves sleeping in the guard-house every night for a week.

The first subject we took up was mechanics. We were made

THE LITTLE CAFÉ ACROSS THE ROAD

to mount and dismount motors, and were familiarized with every part of their construction. Carburettors and magnetos came next, and then we learned what made a motor "go." At the front a pilot always has two "*mécaniciens*" to take care of his machine, but if on account of a breakdown he should have to land in hostile territory he must be able to make the necessary repairs himself, and make them quickly, or else run the risk of being taken prisoner. When flying, the pilot can usually tell by the sound of his motor whether it is running perfectly or not. Many a life has been saved in this way—the pilot knowing in time what was out of order before being forced to land in a forest, on a mountain peak, or in some other equally impossible place.

When we had become "apt," we were promoted to a course in aeroplane construction. This is an extremely technical course, and at first we were asked to know only simple subjects, such as the incidence of the wings, the angle of attack of the cellule, the carrying force of the tail in reference to the size of the propeller. By the incidence of the wings is meant their upward slope. This is an extremely important matter, for the stability and climbing propensities of the machine depend entirely upon their model. The angle of attack of the cellule is the angle of the different wings in reference to each other. For instance, the incidence of one side must be greater than that of the other on account of the rotary movement of the propeller. There are also certain fixed ratios between the upper and lower planes.

Still more important is the carrying power of the tail-plane, for if it has too much incidence it lifts the rear end and makes the machine dive, while if it has too little the reverse happens. If any part of the aeroplane is not correctly regulated it becomes dangerous and difficult for the pilot to control. All this becomes more important as one reaches the close of the apprenticeship. One then appreciates this intimate knowledge acquired at the school. Often a pilot is compelled to land in a field many miles from his base. If something is wrong with his motor he must be able to find out immediately what the trouble is, for if a part is broken the camp must be called up on the telephone, so that a new piece may be sent to the spot by motor, with a mechanic to adjust it.

When a pilot starts on a cross-country trip he is always given blank requisitions, signed by his commanding officer. When he is forced to land he therefore is able to call upon authorities, whether military or civilian, for any service or assistance he may need, and this "scrap of paper" is sufficient in every case to obtain food, lodging, and even transportation to the nearest aviation headquarters for both the pilot and his machine.

Map-reading and navigation were the next subjects we studied. First we were taught how to read a map, how to judge the height of hills and the size of towns, so that when flying we would know at a glance just where we were. This, we later appreciated, is a very important matter. When passing through clouds or mist an aviator may become momentarily lost, and the instant he again sees the ground he must locate on his map the country he views or else land and ask where he is. Aerial navigation may not be as complicated as that employed by mariners upon the high seas, but it is not easily mastered by any means. One must learn to calculate the direction a straight line takes between two points, and translate this direction into degrees on the compass.

Secondly, and more important, is the estimation of the drift caused by the wind. If the wind is from the west and the pilot is attempting to go north, the machine will go "*en crabe*" (sideways like a crab). The machine will be pointing north by the compass, but in reality it will be moving northeast. After the pilot has laid out his course on the map, and is tearing through the air, he must immediately take into consideration his drift. By watching landmarks selected beforehand the drift is calculated very quickly. The course by the compass is altered, and, though the machine is speeding due north, the compass informs the pilot that he is pointing northwest, a fact very confusing for a beginner.

Learning to Fly

During the lecture course we always spent several hours a day on the aviation-field. We were not allowed to fly, but our presence was insisted upon. We would observe the things to avoid, so that when our turn came to go up we should be familiar with all the dangers. Every start, flight, and landing was made a subject of special study. Every time a pupil made a mistake his fault was explained to us, and we were usually impressed with the fact that he had barely escaped a bad smash and perhaps death. The pupils who made the mistakes, were immediately made examples. If the fault was corrected they escaped with a long and loud lecture for the benefit of the onlookers, but if, on the other hand, an accident followed the mistake the offenders were immediately punished with a ten days' "*consigne*." If a pupil continued to make mistakes he was "*vide*," and sent back to his former regiment.

Loss of speed—"*perte de vitesse*," the French call it—is the most common and probably the greatest danger an aviator meets with—it is his "*bête noire*." There is a minimum speed capable of holding an aeroplane in the air which varies inversely with the spread of the wings. While in line of flight, the force of the motor will maintain the speed, but when the motor is shut off and the pilot commences to volplane the force of gravity produces the same result. There are two ways of knowing when you are approaching the danger-point—by closely watching the speed-indicator and by feeling your controls. The moment the controls become lifeless and have no resistance you must act instinctively and regain your momentum, or it is all up with you. While climbing you may lose speed by forcing the motor and climbing too rapidly. When a "*perte de vitesse*" is produced the aeroplane "goes off on the wing," sliding down sideways in such a manner and with such force that the rudders cannot right you or that the propeller cannot pick up your forward speed.

This can happen also if, when making a vertical turn, the speed is not sufficiently increased to carry you around the corner. Occurring near the ground a loss of speed is certain to result in a smash up. If high in the air a "*vrille*," or tail spin, is generally the result. By this is meant coming down in a whirlpool, spinning like a match in the waste of a basin. The machine takes as a pivot the corner of one wing and revolves about it. The first turn is very slow, but the speed increases with each revolution. The only hope of escape is to dive into the centre of the whirlpool. Even then, if the motor is turned on, the planes will fold up like a book. Among the accidents to beginners this, next to faulty landing, is the most common.

I witnessed one very sad example. A young lieutenant had just been brevetted and was ready to leave the school. Just as he was saying goodbye to his comrades, a "*Morane Parasol*" was brought out on the field. These machines are very tricky and dangerous. He had never piloted one, but wanted to show off. The monitors begged him not to take it up, but he insisted on doing so. When he had reached an altitude of about five hundred metres he shut off his motor to come down, not realizing that monoplanes do not volplane well. He did not dive enough and had a "*perte de vitesse*." The machine slipped off the wing. We all held our breaths and prayed that he would recover control before he engaged in the fatal corkscrew spiral. Our hopes were of no avail. The machine started to turn, and approached the earth spinning like a chip caught in a whirlpool. I turned my back, but I could hear the machine whistling through the air till it came to the ground with a sickening crash.

Faulty landings are also very common causes of accidents. It takes a beginner a long time to train his eye to make a perfect

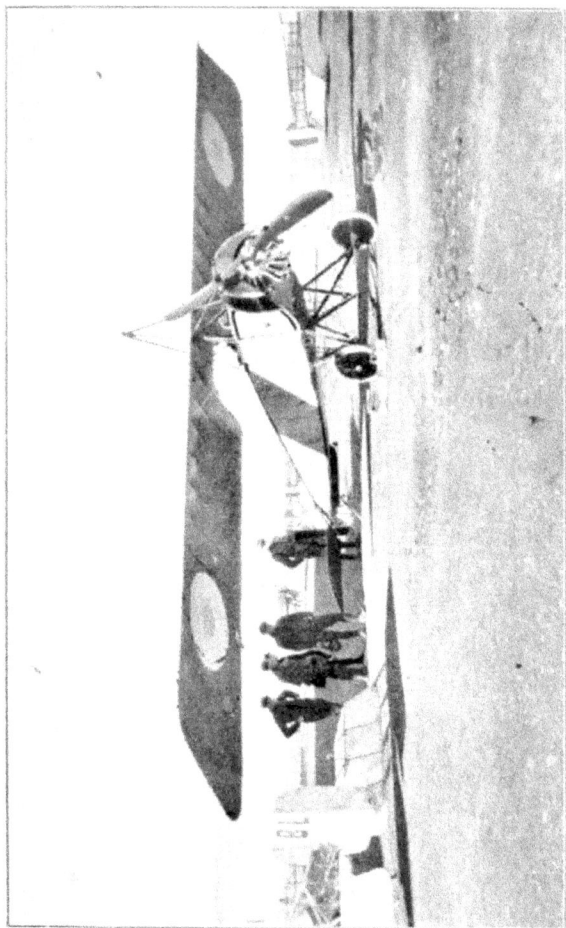

A Morane Parasol

landing, and even experienced men now and then smash up on the "*atterrissage*." A few inches sometimes make a great difference. If the pilot does not check his speed in time he will crash into the ground and "*capote*," that is, turn over. If he pulls up too soon he will slip off on the wing or land so hard that the machine collapses. Not only the manner of landing but gaining the exact place of landing is difficult. If the pilot misjudges his distance, and lands either beyond or short of a given spot, he may collide with some object that will wreck the aeroplane.

Just before leaving the ground is another critical moment. If the tail is lifted too high in an effort to gain speed the wheels are liable to hit some small obstacle and the machine turns a somersault. Often one is forced to lift the tail very high to gain flying speed in a short distance, and it always results in an uncomfortable few seconds until the pilot knows he is clear. Still another mishap against which aviators are powerless may occur while rolling along the ground. The machine may be caught by a cross wind, which will turn it completely around, a "*chevaux-de-bois*," a merry-go-round, the French call it. If the machine is going fast when this happens it means touching the ground with a wing and a first-class smash.

For two months I studied and watched, and the result was a profound respect for the air. During this time it seems that I also had been the object of study and observation on the part of my teachers, for one day I was told that I was to receive my "*baptême de l'air*," my first flight as a passenger. Words cannot describe my joy or my sensations.

I walked over to the double-seater. The pilot had already taken his seat, and the propeller was turning. I had hardly climbed in and fastened my belt than we were off. I could hear the wheels bounding along the ground. Suddenly the noise stopped—we were in the air. I was sure I would have vertigo, as I often had had in high places. I did not look out of the machine until we were about five hundred feet up. Then, to my surprise, I experienced not the slightest sensation of height. The ground seemed to be merely moving slowly under and away from me. We kept climbing. I could see the country for miles. Never had I viewed the horizon from so far. The snow-clad Pyrenees were literally at my feet. Trees looked like weeds and roads like white ribbons. It was a marvellous sight. At about two thousand feet we struck some wind and "*remous*" (whirlpools). Each time we struck one we would drop about fifty feet, and the sensation was like being in a descending express elevator. At the end of the drop we would stop, the biplane would shiver and roll like a ship in a heavy sea, and then it would shoot up until we struck the solid air again. This was real flying.

After a while my instructor cut off the motor, and we started to come down. We were going fast enough on the level, but now the wind just roared past my ears. The ground appeared to be rushing up to meet us. We were pointing down so straight that my whole weight was on my feet, and I was literally standing up. I thought that the pilot had forgotten to redress, and that we would go head first into the ground, but he finally pulled up, and before I knew it we were rolling along the ground at a speed of about forty miles an hour.

With this preliminary experience I was ready to commence my final studies for a pilot's brevet.

Some people seem to think that the two months devoted merely to first principles are time lost, but I now realize that this is not so. I seemed to have much more confidence on account of this intelligent understanding of every detail. I felt that I knew just what I was to avoid, and just how to do the correct thing in case an emergency arose.

Perhaps I might say here that military aviators have four distinct duties to perform at the front: they must fight, reconnoitre the enemy's positions, control the fire of their own batteries, or make distant bombarding raids over the enemy's bases of supplies.

The fighting pilots do nothing but combat work. Their machines are the very small and fast Nieuports, designed especially for quick manoeuvring. They are called the "*appareils de chasse,*" on account of their great speed—over one hundred miles an hour. Their armament consists of a *mitrailleuse*, which is carried in a fixed position. In order to aim it, the pilot must point his machine. The principal task assigned to the Nieuports is to do sentry duty over our own lines, in order to prevent the enemy aeroplanes from crossing over for observation purposes.

Heavier and somewhat slower, and too cumbersome for fighting, are the machines used for reconnoissance duty. They are large bimotor Caudrons, very stable and capable of carrying two men, an observer and a pilot. In addition they carry a wireless apparatus, powerful photographing instruments, and other equipment essential to their work of observing, recording, and reporting the enemy's movements and the disposition of his batteries. If attacked, they can fight, being armed with a machine gun mounted in front of the observer's seat; but attacking the enemy is not the *rôle* they are intended to play.

Next come the Farman artillery machines. They are like the reconnoissance machines, too cumbersome for fighting, but are best equipped for the purpose of *"réglage"*—controlling the fire of batteries. While the small Nieuports have a carrying capacity of only two hundred and twenty pounds, these unwieldy creatures are able to take on over five hundred pounds, which includes the weight of the two men, their clothes, cameras, the wireless, and the gun and its ammunition. In this branch of aviation the weight of the pilot does not matter so very much, whereas in the case of the little Nieuports, if the aviator exceeds the prescribed weight, he has to choose between not piloting the machine or starting on his flight with a supply of gasoline reduced by the amount of his excess weight.

Finally, there are the heavy-armoured bombardment machines, with a carrying capacity of over one thousand pounds. They are the slowest machines of all, and their work is both tiring and tiresome, as their flights are made mostly by night. They are armed with *mitrailleuses* or small, non-recoil cannon, but on account of their low speed their daylight flights are attended by *"escadrilles de chasse."* They are also detailed for guarding cities.

In the early months of the war each aviator was usually assigned to any one of these types of machines at hazard. At the school which he attended the instruction he received was specialized for the work which that particular machine could perform. Since then, however, it has been found more advisable to train all beginners on one of the heavier machines. The reason is this: Fighting, although the safest work, requires the most experienced pilot. It is the most important work of all—or rather it calls for greater attributes of skill, courage, and knowledge. The famous aviators of whom we read in the daily *communiqués*, like Navarre, Nungesser, Vialet, and Guynemer, all gained their reputation with the small, fighting Nieuports. A pilot is consequently promoted from a reconnoissance machine to an "*appareil de chasse*" after he has had two or three months' experience at the front and his captain has indorsed his application.

There are exceptions to this rule. The most notable is the American Escadrille, which consists entirely of fighting machines. The volunteers from the United States who applied for this duty were considered such naturally good aviators that they were accorded the exceptional honour of being assigned immediately to the fighting Nieuports.

When I first reported at the aviation headquarters they offered to let me go directly to the combat school because I was an American. I refused. "*When in Rome, do as the Romans do*," I thought, and so expressed my preference for a French *escadrille*. I knew that by doing so I would put in a longer apprenticeship, but that in the end it would make a much better fighting pilot of me. Having chosen this course, I was ordered to report to the school at Chartres.

The School at Chartres

Most Americans know Chartres only for its beautiful Gothic cathedral, which, since its construction in the eleventh century, has been regarded as one of the finest edifices of France. Those of us who, since the war began, have had occasion to visit Chartres, have found there other interests besides the little, straggling streets, the historic old houses, and the beautiful monuments and memorials to men, like Pasteur and Marceau, who brought fame to their native city in peace as well as in war. Not far from the centre of the town lie the vast aviation-fields—close by the little village of Bretigny, where the treaty of peace which concluded the One Hundred Years' War was signed over five centuries ago. Little did the soldiers who met on that historic battlefield dream that today their descendants would be training for an even greater conflict, in which the combatants not only clinch below ground but also fight aerial battles high above the clouds.

As Chartres was a great cavalry headquarters of the French army before the war, today, (as at time of first publication), many of the horses shipped over from the western plains of North America are sent there the moment they are unloaded from the steamers at Bordeaux. Many a morning from my window I have seen the square below filled with a moving mass of animals on their way to the nearby remount depots. Twice I saw whole regiments of cavalry leaving for the front. So steadily and so quietly did those mounts move in ranks that it seemed as if they had acquired some of their riders' dogged determination and sense of responsibility.

The aviation school at Chartres is as large as the one I had recently left, and its organization is the same. In fact, it is only one of a dozen equally large and important schools located in various parts of France—all of which is bound to make a great impression upon Americans when they appreciate the insufficiency of aviation schools in our own country. At Chartres there are three fields. Two of these are reserved for the use of the double-control machines, while the third, the "*Grande Piste*," serves as a practice-field for the *élèves* who are about to come up for their pilots' examinations. As soon as I had reported my arrival to the officer in charge, and had complied with the usual formalities, I was assigned to an "*équipe*," which in English means a "team." There were twelve of these "*équipes*" under instruction at all times, comprising a dozen pupils apiece, and each having its own double-control machine and an instructor. Half of the number are always making use of one of the two smaller fields, while the remaining six use the other.

The instruction in this way progresses rapidly. The "*moniteur*" first takes each pupil for a few rides to see how the latter takes to the air. The *élève* must follow every move of his pilot, until he appears perfectly at home in the machine, and then he is allowed to hold the controls alone. Each flight lasts only about five minutes, the French theory being that an aviator must take his instruction in small doses. In the interim, as in the preliminary courses, he remains on the field, observing and studying the mistakes of his comrades. If the progress of the instruction is too rapid, the pupil has not the time to grasp each step that has been passed and cannot, therefore, become an expert pilot.

The double control works in this way: The controls and the pedals of the pupil are a duplicate set of those of the "*moniteur*," or instructor, and have the same connections with the engine and steering apparatus. Either set will steer the machine. The pupil takes hold of the controls and places his feet on the pedals. Every motion of the instructor is reproduced in the pupil's control and pedals—their hands and feet move together. In this way the pupil develops a reflex action and instinct for doing the right thing. Each day, weather permitting, at least half a dozen flights are made by a pupil. Gradually the *"moniteur"* allows him to control the machine. Suddenly he finds himself running the biplane alone, with his instructor riding as a passenger behind him and merely giving him a word of advice or caution from time to time.

Landing is the most difficult part of aviation to master. A great many of the accidents occur because the aviator has made a poor contact with the ground. In fact, in the early days of aviation most of the accidents occurred near the ground, and this led people to speculate on the peculiar action of the lower air currents. These, in reality, had little to do with it. The cause lay in the inability of the pilots to know how to make proper contacts and to appreciate the fact which we now know to be a fundamental principle, that the engine should be shut off before a machine catches the air and volplanes down against the wind. There are exceptions to this rule, but not for a beginner. Sometimes it is necessary for the pilot to descend against a strong wind. In order to maintain the required speed the motor must be left partially turned on.

Generally it is most important to turn off the motor, because if the landing is made with the wind, even in the gentlest breeze, the aeroplane, on account of the speed of the tail wind, is likely to turn a somersault and be completely smashed up. Even then another manoeuvre has to be mastered. Just before alighting the pilot must make a quick upward turn, so that at the moment of contact the machine may be travelling parallel with the ground. Formerly the importance of this little upward turn of the rudder was not fully appreciated by aviators, and many a machine was wrecked by a sudden hard compact with the earth.

When the *"moniteur"* sees that his pupil has acquired the knack of making a landing he passes on to the all-important manoeuvre of volplaning, and the dreaded *"perte de vitesse"* is tackled. Lastly, but not least, comes the *"virage"*—turn, or bank, as we say in English. These rudimentary principles are all that are required of the *élève* before he may go up alone, or be "*lâché*."

During this phase of my instruction it was repeatedly impressed upon me that, if anything ever happened to me when I was in the air and I did not immediately realize what to do, I was to let go of the controls, turn off the motor, and let the machine take charge of itself. The modern aeroplane is naturally so stable that, if not interfered with, it will always attempt to right itself before the dreaded "*vrille*" occurs, and fall "*en feuille morte*." Like a leaf dropping in an autumn breeze is what this means, and no other words explain the meaning better.

A curious instance of this happened one day as I was watching the flights and waiting for my turn. I was particularly interested in a machine that had just risen from the "*Grande Piste*." It was acting very peculiarly. Suddenly its motor was heard to stop. Instead of diving it commenced to wabble, indicating a "*perte de vitesse*." It slipped off on the wing and then dove. I watched it intently, expecting it to turn into the dreaded spiral. Instead it began to climb. Then it went off on the wing, righted itself, again slipped off on the wing, volplaned, and went off once more. This extraordinary performance was repeated several times, while each time the machine approached nearer and nearer to the ground. I thought that the pilot would surely be killed. Luck was with him, however, for his slip ceased just as he made contact with the ground, and he settled in a neighbouring field.

It was a very bumpy landing, but the aeroplane was undamaged. The officers rushed to the spot to find out what was the matter. They found the pilot unconscious but otherwise unhurt. Later, in the hospital, he explained that the altitude had affected his heart and that he had fainted. As he felt himself going he remembered his instructions and relinquished the controls, at the same moment stopping his motor. His presence of mind and his luck had saved his life—his luck, I say, for had the machine not righted itself at the moment of touching the ground it would have been inevitably wrecked.

This was a practical demonstration of the expediency of the

French method of instruction, and before long it was to serve me also in good stead.

One day, after I had flown for several hours in the double-control machine, my "*moniteur*" told me that he thought me qualified to be "*lâché*," and that I was to go up alone the following morning. I felt very proud and confided my feelings to one of my friends who had been qualified a few hours earlier. While we were talking he was called upon to make his first independent flight. We watched him leave the ground, rise, and then make his turns. He was doing remarkably well for a beginner, but when he came down for his landing he did not redress his machine in time and it crushed him to the ground, with fatal result.

This completely unnerved me. I lost all desire to fly the following day, and prayed earnestly for rain. The next morning, however, was beautifully clear. The captain was there to watch my flight. I was loath to go up, but I had no alternative. The mechanics rolled out a single-control biplane for my use and I climbed in. The motor was started. With its crackling noise my nerve almost deserted me again. I should have felt less frightened, probably, had no one been looking on, but my "*moniteur*," my captain, and all my comrades stood there, interested to see how I would handle myself. I had to see the thing through, so I opened the throttle.

The machine began to roll along the ground, then to bounce, and then, in response to a pull on the control, to fly. I was flying alone. The thought filled me with alarm. I rose to less than two hundred feet, but it seemed prodigious. Then I made a turn. When I found that I was flying smoothly and easily I felt a little more confident.

As I turned back toward the field I could see my masters and comrades below looking up at me. Another machine was about to leave the field. It seemed no larger than a huge insect as it glided across the ground. I made up my mind that I was going to make good. If others could do it, I could. I volplaned down, and made my landing safely but somewhat bumpily. The captain told me that I would do, but he would like me to make another turn. I went up again. This time I made a faultless landing. I had passed my test satisfactorily. I felt happy and confident. I was now qualified to "conduct" an aeroplane alone, and in a few weeks I would be allowed to try for a brevet as military pilot.

There were several other pilots whose turns to pass to the "*Grande Piste*" came before mine. I had, therefore, to wait for several days, which I used to advantage in taking up the old "double-control" machine alone. In this way I was able to make several ascensions and landings every morning and every night. This was to be of the greatest service to me later, for during these practice flights I acquired perfect confidence in myself. At other times, both before and after working hours, my "*moniteur*" would take me up with him as a passenger for a newly discovered sport.

We would rush along the ground, barely two feet above it, and put up partridges, which abounded in the greatest numbers. Our speed would enable us to overtake and hit them with the wires of the machine and kill them. Running along the ground in this way is always attended with danger, but it was real sport. One morning in twenty minutes we killed six partridges in this novel manner.

Finally my turn came. I graduated from the beginners' class at "*La Mare de Grenouille*" to the company of the more finished pilots of the "*Grande Piste*." The beginners' field is called the "frog's meadow," because the landings are so hoppy. On the "*Grande Piste*" we had newer and faster machines, and we could fly alone and go practically anywhere we wished.

Six pilots were assigned to one aeroplane. We had to divide up the time equally between each pilot, so as to give everyone an opportunity of making at least two flights both morning and afternoon. A maximum height was imposed upon each "*équipe*," and this was gradually increased from five hundred feet to a thousand, and then to one thousand five hundred, as we became more and more adept.

Most of our time was given to making landings and to accustoming ourselves to volplaning. The motor had to be reduced at a predetermined distance from the field, and the rest of the descent made by volplaning to a given spot. Spirals were also made during each flight. We would select our landing-places and prepare ourselves for the "*atterrissage*" by reducing our motor, making due allowance for the drive of the wind. At about two hundred feet from the ground we would suddenly turn on the motor again, tilt up the tail, and resume our flights. This was excellent practice and gave us more and more confidence in our own ability to come down wherever we wished. The average layman cannot understand why aviators spend so much time turning in spirals as they approach the ground. It is because they are manoeuvring for position to hold their headway and land against the wind, as does a sailing ship when beating up a harbour against wind and tide.

Every day I took my machine up higher and higher until I had gradually increased my altitude to two thousand feet. Here, one day, I had a narrow escape. I had received orders to make a flight during a snow-storm. I rose to the prescribed height and then prepared to make my descent. A whirling squall caught me in the act of making a spiral. I felt the tail of my machine go down and the nose point up. I had a classical "*perte de vitesse*." I looked out and saw that I was less than eight hundred feet above the ground, and approaching it at an alarming rate of speed. I had already shut off the motor for the spiral, and turning it on, I knew, would not help me in the least.

Suddenly I remembered the pilot who fainted. I let go of everything, and with a sickening feeling I looked down at the up-rushing ground. At that instant I felt the machine give a lurch and right itself. I grabbed the controls, turned on the motor and resumed my line of flight only two hundred feet in the air. All this happened in a few seconds, but my helplessness seemed to have lasted for hours. I had had a very close call—not as close as the man who fainted, but sufficiently so for me.

Since that day I have seen several other pilots experience a loss of speed under similar circumstances. Thanks to the thorough instruction which we had received previous to our being allowed to fly alone, their lives, as well as my own, were saved. Later we learned how the very dangers which we had experienced as new aviators often become the safety of expert pilots.

I HAD RECEIVED ORDERS TO MAKE A FLIGHT
DURING A SNOW-STORM

THE AUTHOR, TOGETHER WITH HIS FIRST MECHANIC, AT THE
"*MITRAILLEUSE.*"
THE SECOND MECHANIC IS STANDING ON THE WING.

Passing the Final Tests

My *équipe* was now making flights at three thousand feet and was remaining up for an hour at a time. We had all flown alone for thirty hours and were ready for our "*épreuves*."

The weather was cloudy, however, and as our first examination was to be a height test, we had to wait until it cleared. It would have been extremely difficult—in fact, almost impossible—for us to go up under existing conditions. The first two tests which we were required to pass involved ascensions of six thousand feet; then an hour at ten thousand. If we passed these satisfactorily we would next be required to take a triangular voyage of one hundred and fifty miles, making a landing at each corner of the triangle. Lastly, there was the ordeal of going up to an altitude of one thousand five hundred feet, where the motor had to be cut off and the descent made by spirals to a previously determined spot.

The day on which we were required to begin our altitude flights the captain assigned three machines to our *équipe*—that is, one aero-biplane for each pair. My chum, a *sous*-lieutenant, and I were assigned to the same machine. We matched to see which one of us should use it first. He won and I helped him prepare for the test. I fastened on his recording barometer, which indicates the altitude reached by a machine, and he climbed in. Waving us a cheery "*Au revoir*," he started off. His machine climbed fast. To us he seemed to be going too steeply. We felt like shouting to him to be careful, but we knew it was useless.

Suddenly his machine slipped off on the wing. For some unknown reason he failed to shut off his motor. His biplane engaged in the fatal spiral. There was a loud report, like a cannon-shot, and the machine collapsed. The strain had been too great. The top plane fell one way, the lower another, while my friend and the motor dropped like stones.

I would have given anything to put off my own test for a few days, but within twenty-four hours I received orders that my turn had come; and orders were orders. I made up my mind to be very careful and to take my time about the climb.

That first flight at six thousand feet gave me a thrilling sensation. I remembered my first flight alone, when I had barely reached two hundred feet. It seemed now as if I was going to mount to an indescribable height. Since that day I have had to go up that high often, and even higher; but it has all become commonplace, for *familiarity breeds contempt* in the air as well as on land.

I was so very cautious about mounting that twenty minutes elapsed before the needle on my registering barometer marked six thousand feet. It was very cold. The wind struck my face with icy blasts, but I was so excited that I did not really mind it. After a while I shut off the motor and started to volplane to earth. I came down a little too rapidly and made a very bad landing. In fact, for a moment I thought that I had broken my machine. I was wet all through from the sudden rise in temperature and stone-deaf. It was ten minutes before I could hear again. Then I received my call-down. It seems that when a pilot has been up to a very great height, he loses his sense of altitude—his "*sens de profondeur*," as the French call it. When approaching the ground he cannot tell whether he is twenty-five or fifty feet in the air. He must take every care, before making contact, to train his eye for "depths" again by flying a few minutes fairly close to the ground. It is only a question of a few moments, but it is a necessary precaution.

My next climb to six thousand feet was better. In fact, I felt a certain degree of confidence. It took me somewhat longer to mount to the required height, because some clouds came up and

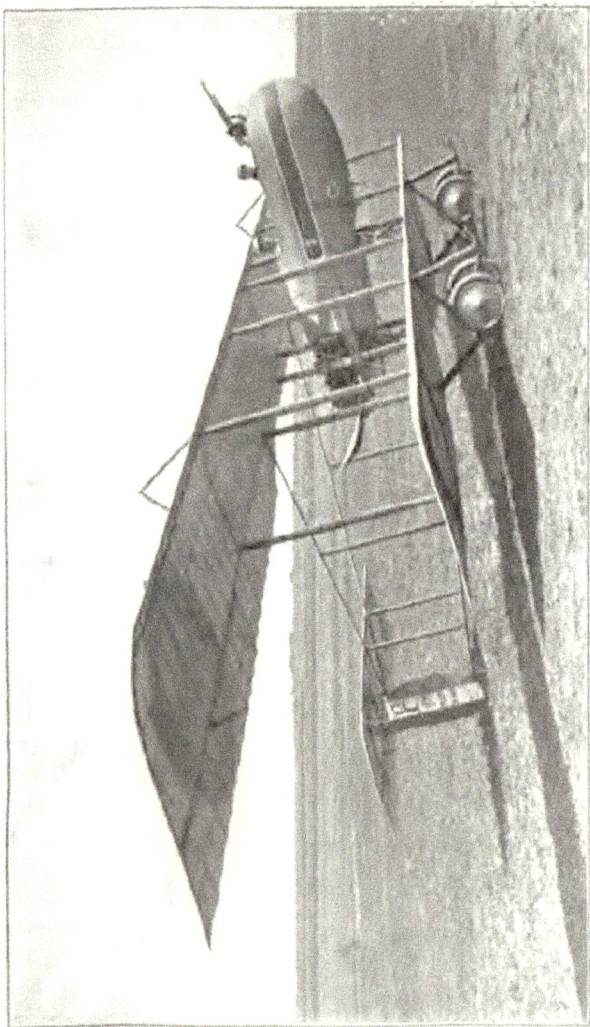

A FARMAN ARTILLERY-MACHINE

I had to search for a hole through which to pass. Everywhere below me, as far as my eyes could reach, was a sea of clouds. The sun was shining on their snowy-white crests. It looked for all the world as though I was looking down upon an enormous bowl of froth.

The following morning was the day fixed for my ten-thousand-foot ascension. The atmosphere was remarkably clear, and I felt an extraordinary sense of freedom and power as I rose from the ground. The earth below me was bright with colour. As I climbed higher the shades became less brilliant. At ten thousand feet all colour had vanished. The only hue visible was a varying degree of shading, gray and black. Below me I could make out the city, of Chartres. Forty miles away lay Orleans. To one side, the Loire wound its course in a gray, ribbon-like band. On all sides the straight, white roads were merely blurred streaks in the murky mass.

A few days later I started on my endurance test, the triangle, in company with five other machines. In this flight of two hundred and fifty kilometres I had to make landings at two towns where there were aviation-fields, and the third my own field. At each place I had to report to the aviation officer in charge and have my papers signed. In case of a breakdown on this flight I had forty-eight hours in which to make the necessary repairs and complete the test.

The day of my triangle was a poor one for flying. It was the first warm morning of early spring and the sun was just soaking the moisture out of the ground. The air was, in consequence, spotty and there were many "*remous*," or whirlpools. These whirlpools often cause a sudden "*perte de vitesse*" and are therefore very dangerous. The machine is sailing along quietly and smoothly, when suddenly the controls become lifeless. You glance at your speed-indicator and at your engine-speed. Both show that the machine is travelling well above the minimum safety speed. This is apt to puzzle the beginner, for without warning there follows a sudden jolt. Your machine trembles like a frightened horse and unexpectedly leaps forward again. On a

45

day like this you have to fight the machine all the time to maintain its equilibrium.

The first leg of the triangle was accomplished without incident. As I was starting my motor for the second stage, however, I noticed that the ignition was faulty. A spark-plug had become fouled with oil, and I had to change it before venturing up again. My companions started without me, calling out that I could catch up with them at Versailles, where they intended to lunch. I hurriedly screwed on a new spark-plug and threw my tool-bag back into the box under the extra seat, but in my hurry to be off I neglected to fasten it down. I was later to regret my carelessness.

I soon found that in trying to catch up with the others I had no easy task before me. The day was well advanced, and the "*ramous*" which I encountered were countless. I climbed and climbed. To no avail. The cloud ceiling was at eighteen hundred metres, and I could not escape the "*remous*" so low. The country below me was all wooded and interspersed with lakes both large and small. There was not a landing-place in view. Suddenly I felt a hard blow on the back of my head and a weight pushing against me. "*Ça y est*," I thought; "the machine can't stand the buffeting and has given way." I ventured a look back. To my surprise, everything seemed intact—everything except the observer's seat, which was leaning against my head! It was the seat which I had forgotten to hook down at Châteaudun! I was greatly relieved, and fastened it back into place.

Just then I came within sight of Versailles. I looked for the aviation-field at which I was supposed to land. Instead of one I saw three, lying about two miles apart. This was indeed a puzzle. From the height at which I was flying I could not make out which was the school. I picked out one which I thought should probably be the haven of refuge for my storm-tossed aeroplane and spiralled down. I climbed out of my machine. No one seemed to be about. No mechanics ran out to assist me, as is usually the case at the schools. "It must be the luncheon-hour," I thought, "and all the *mécaniciens* are at *déjeuner*." I glanced over

46

to where the machines were ranged in line.

To my surprise, they were not of the model I had seen at Pau and Chartres, but the latest and fastest "*avions de chasse*." Somewhat uncertain as to my whereabouts, I walked over to the office. I was not left long in ignorance of my error. I had landed on a secret testing-field, access to which was obtained only by special permit. The sergeant advised me to lose no time in leaving, for if the captain saw me I would be speedily punished in accordance with the military regulations. I needed no second urging, and within five minutes I was on the right field, explaining to my comrades why I had been so long rejoining them. It seems that they had experienced a very pleasant flight all the way, for the hour's start they had had over me had enabled them to escape most of the "*remous*," which are always at their worst in the middle of the day.

Late in the afternoon we returned to Chartres. This was the most enjoyable part of the day's flying. The aerial conditions were perfect and we were able to allow ourselves the pleasure of appreciating all the interesting places we passed over. First we saw the beautiful valley of the Chevreuse; then Rambouillet, with its wonderful hunting and fishing preserves. Next I caught a glimpse of the imposing palace and gardens of Maintenon. The time passed all too quickly; yet when we reached home it was almost dark. We all felt quite tired, but before putting our machines away, however, we asked permission to make our spirals, so that we might complete every requirement of the brevet before night set in. We were anxious to do this, so that we might obtain our "permissions" immediately. We did not wish to lose a moment. A four days' leave is always accorded each pilot the moment he has satisfactorily fulfilled all the requirements of the course. Our request was granted and the final test was successfully passed.

I was now a full-fledged aviator, with the rank of corporal, with the regular pay of eight cents a day and an additional indemnity of forty-five cents as a member of the Flying Corps.

The Zeppelin Raid Over Paris

I decided to spend my four days' "permission" in Paris, the rendezvous of all aviators when not on active service. From the first I felt conscious of unusual attention. People seemed to treat me with deference and with more respect than I had ever before experienced. I could not account for it. Then, of a sudden, I chuckled to myself. The envied stars and wings on my collar were the cause. I was a "*pilote aviateur*," a full-fledged member of the aerial light cavalry of France.

For most "*permissionnaires*" Paris usually offers only the distractions of its theatres and restaurants, its boulevards, and its beautiful monuments. These pleasures I also had looked forward to, but in the first thirty-six hours of my visit occurred another, more startling diversion—two Zeppelin raids. It was my first real experience of the war.

The first alarm occurred as we were leaving a restaurant after dinner. A motor fire-engine rushed by, sounding the alert for the approaching enemy. Pandemonium reigned in the streets. I hastened to find a way to reach the aviation-field at Le Bourget, where I felt that duty called me. The *concierge* hailed a taxi. I jumped in and gave the address to the chauffeur. "*Le Bourget! Oh, mais non*," exclaimed the man; "*monsieur* must think me a fool." He flatly refused me as a fare. He was the father of a family, and he certainly would not go to the very spot where all the bombs were certain to be dropped; besides—he did not have enough gasoline in his tank for so long a run. We talked and argued. In desperation I thrust my hand in my pocket and handed

NEWSPAPER DROPPED BY GERMAN RAIDERS
WITHIN THE FRENCH LINES.

him a generous retainer. At the sight of the money he wavered. I followed up my advantage and promised him a handsome tip if he started at once. He threw in his clutch. I had won my first "engagement."

The streets were pitch-dark and jammed with people, all staring heavenward. The feeble oil-lights of the taxicab barely lit up their faces as we wound our way in and out. At break-neck speed we swung right and left, sounding the horn and crying out warning "attentions." Near the outskirts of the city we could see searchlights flashing against the heavy mist. There was so much fog, however, that they could not pierce the veil which hung over the city. At one thousand five hundred feet the sky was opaque. The anti-aircraft batteries were barking and sending off deep-red flashes into the impenetrable murkiness in answer to wireless signals from the invisible air guards above. Now and then a military automobile dashed by.

As we neared Le Bourget, there was a deep detonation. A bomb had been dropped. The Zeppelin had arrived. My chauffeur in panic jammed on the brakes. I was literally thrown out of the taxi and into the arms of a waiting sentinel who flashed an electric torch into my face. The sergeant of the guard rushed up and escorted me to the guard-house, where an officer proceeded to question me. I immediately realized that I was an object of suspicion. Who was I and what was I doing here? Here I was, a foreigner in the French uniform, and unknown to them. Instead of being welcomed at the post of danger I found to my amusement that I was temporarily under arrest.

Several more explosions were heard. Then a deathlike silence. The cannon ceased their angry roar, the searchlights put out their blinding rays. Through the window I noticed a large fire in the middle of the "*piste*," where several cans of gasoline had been ignited. It was the signal for the searching aeroplanes to return. The Zeppelin had left.

As soon as the electric current had been switched on again the captain returned. He seemed surprised at my "enthusiasm." "Just like you Americans," he said smiling. "A man *en permis-*

sion, however, should never look for trouble." He then explained that this night guarding required special training. Even had he needed my services, I would have been helpless, as I had never before flown after sundown.

One by one the defenders of Paris returned. At two thousand feet they were invisible, though we could hear the humming of their motors. Then, as they came nearer and nearer we saw little indefinite lights moving in the mist above us, and finally the machines, their dimmed searchlights yet staring like two great eyes.

About fifty aeroplanes are in the air around Paris all the time. Each pilot remains up three hours, when he is relieved by another flier. When the Zeppelins are known to have crossed the front, some eighty miles away, the whole defence squadron of two hundred takes to the air. The organization of the aerial defence of Paris is admirable, and it is this, together with the efficient anti-aircraft posts in the environs, which prevents the "Boches" from raiding Paris more often.

I was allowed to examine everything at my leisure, and took advantage of this opportunity to gain as much information as I could about the lighting systems and the new models of small cannon which had recently been installed on the aeroplanes. Presently I was greeted by one of the pilots who had just landed. He proved to be an old acquaintance.

It seems that the Zeppelin had profited by the mist to slip by our watchers at the front, and had reached the very outskirts of the city before it was sighted by the air guards of Paris. The Boches dropped several bombs near the Gare du Nord and in the vicinity of Le Bourget. Then they had vanished into the mist.

"How could they ever find the railway-station in the dark?" I asked.

"That's easy," he answered. "The Zeppelins are equipped with a small observation-car that hangs down on a long cable. It is built something like an aeroplane and travels about five thousand feet below the dirigible. This evening the raider flew at an altitude of seven thousand feet, while the car moved along

An anti-aircraft .75

only two thousand feet from the ground. Its passenger could, therefore, locate everything easily and telephone the directions to the commander above."

"But," I insisted, "how did they ever locate the freight-yards in the dark?"

"Easily," replied my companion; "their spies had arranged all that. They simply hung a series of blue lights in the chimneys of houses and laid out a path directly to the spot."

These spies in all probability had been already caught, but I was angry, very angry, to realize that their "espionage" was still so efficient.

On the way home Paris seemed surprisingly normal again. The street-lamps were glowing peaceably and the cafes were crowded with talkative men and women. I could not help thinking how wonderful those people were, how fearless and forgetful of danger.

The actual damage done by the bombs during that raid was insignificant. The photographs published in the daily press bore witness to this. A few civilians were killed, but no military damage was done. It was only an attempt at terrorism. I visited one of the "craters" the next morning. The bomb had landed directly over the subway and had blown a huge hole in the pavement. The tracks below lay open to view. Gangs of labourers were already at work, not repairing the damage but enlarging the hole. I asked them what this was for. "Why, *monsieur*, it is this way. The health authorities always insisted that a ventilator was needed in this part of the '*Metropolitain*.' The Boches obligingly saved us the trouble and expense. We are now merely going to put a fence around it."

That night there was another alarm. We were spending the evening with friends in the Latin Quarter when the *pompiers* startled us with their wailing sirens. From every direction came the "*Alerte!* the Zeppelins are coming. Lights out!" One by one the street-lamps faded, apartments were darkened, and the street-cars stopped where they were, plunged into darkness. It was thrilling. In the velvety gloom the outlines of people and motors could be seen moving about. The corner of the *rue d'Assas* alone remained illuminated. A "*bec de gaz*" was still burning brightly, to the rage of an old infantry colonel who was too short to reach it himself. To our amusement, a little girl clad in a red *kimono* and bedroom slippers ran out into the street and volunteered her aid. The old soldier blurted out a word or two, then lifted her up in his arms while she extinguished the light. "Thanks, *mademoiselle*. Now, quick!" he gasped; "run back to bed."

We saw some of the people climb down into their cellars. The majority, however, gathered in the streets, looking up at the searchlight swept sky. Tiny, starlike lights moved about above us and we knew that aeroplanes of the "*Garde de Paris*" were searching for the venturesome raider. "I don't believe the sales Boches and their sausage balloon are coming this evening to beg food," remarked one man.

"Oh, no," answered another, "it is clear and they well know that a Zeppelin over Paris tonight is a Zeppelin less for Germany."

Just then we heard the firemen coming back. Their bugles were playing a jubilant call. The Zeppelins had been frightened away. Everywhere lights were again lit. The people laughed good-naturedly at their neighbours' strange attire.

"*Quelle guerre!*" yawned the old officer at my elbow; "down in the ground, under the sea, and over our homes! *Quelle guerre!*"

At the Ecole de Perfectionnement

From Chartres I was sent to Châteauroux to continue my studies and perfect myself in flying. Châteauroux is a small provincial town situated half-way between the *château* country and the beautiful valley of the River Creuse. It was originally founded by the Romans, and before the war had a large "*caserne*." All this is forgotten today in the glory of the stream of air pilots that pass through the "*Ecole d'aviation militaire*." Soldiers are to be found everywhere, but not aviators, and the residents of Châteauroux are very conscious of the honour conferred upon their town.

When a pilot has received his "brevet" he has really only begun his professional education. This I soon found out at Châteauroux. The day after my arrival I was set to work making daily flights and attending the various courses and lectures on artillery-fire, bomb-dropping, war aviation, "liaison," and the design of enemy aircraft. The daily flights were very short, lasting only fifteen minutes each. We made three or four of them each day, and their purpose was chiefly to give us greater confidence in making our landings. We were allowed to take up passengers, and we often paired off and took each other up.

In this connection it was amusing to see how everyone avoided being taken up by certain pilots. Some men cannot fly: their temperaments prevent it, and try as they will they cannot improve. This is generally due to sheer stupidity or to lack of nerve. One thing is certain, and that is that these men will kill themselves sooner or later if they persist in their efforts to fly.

An incident occurred shortly after my arrival at the school. About thirty pilots were receiving practical instruction on the aviation-field and were standing around two aeroplanes. About a hundred feet away another machine was making ready to start. When the mechanic spun the propeller at the word from the aviator the motor started, not slowly as it should, but with a roar. The machine began to roll toward the group of men. Instead of cutting off the ignition—we found out later that the wire connecting the throttle and the carburettor was broken and that the throttle was therefore turned on full—the pilot lost his head. He tried to steer around the group of men in front of him. The ground was muddy and very slippery, which made escape almost impossible. In their hurry to get away several men lost their footing and fell down in the very path of the onrushing biplane.

We thought that at least a dozen would be crashed or else decapitated by the rapidly revolving propeller. Fortunately no one was seriously injured. Even the stupid pilot escaped unscathed. The three machines, however, were completely wrecked. Needless to say, the offender was immediately dismissed from the aviation school and sent back to his regiment. His escapade had cost the government about ten thousand dollars. Even had there been no damage to the machines it is doubtful whether any further chances would have been taken with a man of such a temperament.

There is not much to tell of the daily flights which we made. The weekly trips, however, proved extremely interesting. We usually covered at least a hundred miles and flew at a height of over six thousand feet.

My first trip was to the aviation school near Bourges, situated on the estates of the Count d'Avord, who has lent the ground to the government for the duration of the war. It is a much larger school than any I had attended, and its instruction covers every type of machine. The most important course given is that in night flying. All the aviators who have been selected for the bombarding-machines and for the work of guarding cities are

A BAD LANDING

sent here. Their life is the exact opposite of that led by the average pilot, for they sleep all day and work all night. All this was so new to me that I found much of interest.

What surprised me most was to learn that night flying is really easier than day work. The reason given is that after sunset there are no "*remoux*," and that, when it comes to making landings, the aviation-fields are so well lighted that the pilots have no more difficulty making contact than in the daytime. It is another matter, of course, if an aviator meets with a mishap and is forced to alight elsewhere. Under those circumstances the story is usually a sad one.

One of the longest flights I made was to Tours. Since then I have often thought how strange it was to be flying over this historic region of France. We took it as a matter of course, but what would the ancient heroes of France have thought had they seen us? One week we were over Bourges, called the source of the French nation, for it was from here that the Duke de Berry sallied forth and conquered the English hosts, bringing to a close the struggle which had lasted for a century. The next week we were over Lorraine and the *châteaux* of the Bourbon kings, who did many great deeds, but had surely never thought of flying.

The lectures which we attended every day were extremely important. The first subject covered had reference to artillery-fire and the theory of trajectories. It is essential that aviators be familiar with the parabola described by the shells fired by cannon of various calibres. If they are not, some day they may unconsciously fly in the very path of shells sent by their own guns and be killed by projectiles not meant for them. The "seventy-five" field-guns, when firing at long ranges, have to elevate their muzzles so much that their shells describe a high parabola before they explode over the enemy's trenches.

The very heavy shells, on the other hand, like those of the 420-centimetre French pieces and the famous German "Big-Berthas," rise to a point almost over their target and then drop suddenly. Aviators must become familiar with this and with a hundred other peculiarities of artillery-fire. When flying over the front it is too late to acquire this knowledge. Information has to be gained beforehand or you stand the chance of being annihilated with your machine. An aviator I know involuntarily got into the path of a seventy-seven or seventy-five calibre shell. He is alive today, but he lost his left foot in the "collision." He just managed to come down within our lines before he had bled too much to recover.

Our next subject for study was "liaison," which means the science of maintaining communication between the several branches of an army. During an attack upon the enemy's position each arm of the service has its own part to play. The artillery has to prepare the way for the infantry, and at a given signal the infantry must be ready to rush forward to the attack. As the infantry carry the positions before them and move forward the artillery-fire must be correspondingly lengthened; the supply-trains have to keep the necessary amount of ammunition and shells and other material supplied to the infantry and artillery; while the cavalry must be ready to charge the moment a favourable opening presents itself.

For all this co-ordination there are various *"agents de liaison."* There are messengers on foot, and despatch-bearers on horseback, and motorcycles; there are visual signals, such as the signal-flags, the semaphore, and the coloured fires and star shells; and there are the telephone and the telegraph. None of these are depended upon by the army headquarters as much as the aviation corps, the *"agent de liaison par excellence."* It is one of the most important *rôles* that aviators are called upon to play at the front, and we were being prepared for this work by very special instruction.

Under the subject of "war aviation" we studied the designs of the various enemy aircraft, and the pitfalls which are encountered at the front. Then followed a course in bomb-dropping. This was a practical course, and our method of learning was as peculiar as it was ingenious. A complete bomb-dropping apparatus was mounted on stakes about twenty-five feet above the ground. Under this there was a miniature landscape painted to scale on canvas. It was a regular piece of theatrical scenery mounted on rollers so that it could be revolved to represent the passage of the earth under your machine. We would climb into the seat on the stilts and consider ourselves flying at some arbitrary height. Through our range-finder we would gaze down at the "land," and as a town appeared we would make allowances through the system of mirrors arranged by the range-finder for our speed and height and for an imaginary wind. At the calculated moment the property bombs would be loosed.

When I came to Châteauroux I thought that I knew something about aviation because I had obtained my "brevet." I soon realized how very little of the ground I had actually covered. In fact, after four weeks of this advanced work I felt as if I would never acquire all the knowledge required for work at the front. Just then about twenty of us were selected to go into the reserve near the front, to fill vacancies caused by casualties. At last! We were off for the front!

I left Châteauroux for the reserve at Plessis-le-Belleville with a certain feeling of uneasiness, yet with the certainty that in case of emergency I knew almost instinctively what to do. In addition, I had become thoroughly familiar with the perils of the air which pilots are called upon to meet most often. These dangers are the same as those encountered on the high seas by sailors: fog, fire, and a lee shore. Take fog, for example. The most difficult operation in flying is the "*atterrissage*" (landing).

Now, in a fog you must land almost by chance. You cannot see the ground until it is too late for your eyesight to be of any use. Your *altimetre* is supposed to register your height above the ground, but no *altimetre* is delicate enough to keep up with the rapid descent of an aeroplane. It is always from fifteen to twenty yards behind your real height. Nor is this all. The *altimetre* "begins at the ground"; it registers your height above the altitude from which you started. Now, since all ground is more or less irregular, you may be coming down on a spot lower—or, much worse, higher—than that from which you started. Besides, when a fog comes up the atmospheric pressure changes and, as the *alti-*

metre is a barometer, it becomes from that moment unregulated. Then, of course, on landing you may strike bad ground—houses or shrubbery or fences, all of which adds to the uncertainty and risk.

The danger from fire has never been entirely eliminated, although it is not today, (as at time of first publication), as great as it was before aeroplane engines reached their present perfection. The greatest danger lies in the propeller. The slightest obstacle will break it, and if the motor cannot be stopped instantly the increased revolutions are certain to force the flame back into the carburettor and you are "*grillé*" before you can land. Aviators are from the first instructed to leave nothing loose about the machine or their clothing. Many pilots have been killed because their caps blew off, caught in the propeller, and broke it. So fast and powerful is the motion of the propeller that I have seen machines come out of a hailstorm with the blades all splintered from striking the hailstones. There have been experiments made with fire-proof machines, but none have yet proved successful. Fireproofing is apt to make a machine too heavy and cumbersome.

The last peril of the air we were warned against was that of the lee shore. In landing you should always do so against the wind. This is the first principle drummed into the beginners at the schools. If you make an "*atterrissage*" with the wind behind you, you roll along the ground so fast and so far that you are apt to meet an obstacle which will either wreck your machine or else cause it to turn a somersault. Yet, when making a landing against the wind, the force of the breeze blowing toward you will sometimes prevent you from coming down where you had planned. On many occasions I have seen aeroplanes remain practically stationary in the air, while descending, and sometimes even move backward in reference to the ground. This has to be considered by the pilot and grasped on the instant, or else he will surely come to grief by hooking some object in his descent.

The Réserve Générale de L'aviation

The "Réserve Générale de l'Aviation," "R. G. A.," or, as it is more commonly known, the "*Groupe d'Entraînement*," is situated northeast of Paris, on the plains of Valois. It was there that General Maunoury, in September, 1914, launched his turning movement against the German right flank under General von Kluck, and helped save France in the great battle of the Marne. The country in this section is ideal for aviation, for the hills are low and rolling, and there are very few "obstacles." In a large forest— an "obstacle"—the village of Ermenonville lies. Here we were billeted, while the commanding officer of the reserve made his headquarters in the *château* of the Prince Radzivill, the "*patron*" of the neighbourhood.

The organization of the reserve is stupendous. There are four separate camps, one for each branch of aviation, and there are over one hundred machines in each camp. We were practically our own masters, and could make flights whenever and wherever we wished. The idea is that the pilots here have an opportunity of perfecting themselves and that, if they do not fly, why, then it is their own loss. Acrobatics and all sorts of feats are encouraged. Accidents occur every day, but we were here on "active service" and our time was far too taken up with our work for anyone to pay much attention to the unlucky ones. That, at the front, is a duty reserved for the medical corps.

Now that we were all gathered in one great camp, I had the opportunity of noticing more than before the different types of men that are to be found in the French Flying Corps. Unlike

65

the conscript *"poilu"* of the army, every aviator is a volunteer. Aviation is far more dangerous than fighting in the trenches, yet there are many who have preferred the extra risk of being in the Aviation Corps to the tedium of remaining in the narrow-walled trenches. I believe there is at present a waiting list of over six thousand men who have applied for service in the Flying Corps, but for whom there are still no vacancies. A pilot may resign his commission at any time and return to his regiment at the front, but the majority of the "vacancies" are caused by casualties. Curiously enough, there are many men who have been rendered unfit by wounds for service in the infantry, who have volunteered for the air service. A people with such patriots surely can never be defeated.

The French army understands that flying calls for the most intense kind of concentration, mental as well as physical. Every effort is therefore made to absolve the aviators from all work except that of running the machines and seeing that they are well cared for. My old football trainer in college used to say that his principle was to wrap the men in cotton-wool when they were off the field and drive them for all they were worth when they were in the field. The French seem to have the same theory about aviation. No one who has not tried it can appreciate the tremendous strain of flying. After a few hours in the air I find that I am more exhausted than I used to be after a hard football match.

While the matter of personal habits is left to the aviator's judgment, he is usually cautioned about drinking, smoking, and even overeating. You need all the strength there is in you when in the air. The French, as everyone knows, drink wine as we drink tea and coffee. Yet I have noticed that French aviators, when they are at work at the front, merely colour their water with the wine. Many of them smoke cigarettes only in moderation.

The democracy existing in the French army since the outbreak of hostilities has aroused the enthusiasm of every observer and has caused much surprise to incredulous pacifists. The Aviation Corps I found even more democratic among themselves than the other branches of the service. I suppose one of the reasons for this total absence of distinction between officers and men is because they all have passed through the same schools, through the same courses of training, and have run the same risks.

Among the pilots, however, one may notice three classes. The first and predominating class is that composed of "gentlemen." By gentlemen I mean gentlemen in the English sense—men who in private life have the leisure time to be sportsmen, and who in war have chosen aviation because it is a more sporting proposition than fighting in the trenches. The second class comprises those who before the war were professional pilots or aviation mechanics. In the third class one finds men who were mechanics or chauffeurs by trade and who were accepted because their knowledge of machinery would ultimately help them to become pilots.

The best pilots are obtained from men between the ages of twenty and thirty. Under twenty a boy is too impetuous, and over thirty a man is apt to be too cautious. Of course, there are exceptions, but these limits express the preferences of the instructors at the schools.

At the "R. G. A." there is also a course for the training of young artillery officers who have volunteered as observers for the aviation. Our duty as pilots was to take one of these officers up with us every time we made a flight, so as to give him "air sense." We would make imaginary reconnoissances all over the country, regulate supposed batteries, and go on photographic missions. The observer would send off his reports by wireless and direct us by his maps, while we would do our best to throw him off his guard and make him lose his bearings. In this way observer and pilot work together and help each other with observations and advice.

During my stay at the "R. G. A." my partner was a young artillery officer who had just been promoted from the ranks. He was very clever and full of enthusiasm for his work, and we derived much pleasure from our association.

"*Réglage*," or fire-control, was a course that involved practice, constant practice, and still more practice, in developing a faculty for reading distances. We would go up and try to estimate just where the puffs of smoke, representing the explosion of shells, went off. The corrections then had to be wirelessed to the battery, so that the next shots might get "home." In real observation work the observer does all this. The pilot merely flies the machine. It is thought best, however, for the pilot to have the same training and technic as the observer, so that he may help the latter.

While stationed at the Reserve I made some most interesting trips around the country, up and down the valley of the Marne, over the forest of Compiègne, and even over Paris. In fact, I was at liberty to go anywhere, except in a north-easterly direction, for there there was always a danger of getting across the lines. Two or three machines disappeared in the course of a year, and it is thought that the pilots must have committed an "indiscretion" and fallen into the hands of the Germans for their pains.

Once I was sent to Bar-le-Duc to bring an old machine back to Plessis. The distance, as the crow flies, was one hundred and fifty miles, over Châlons, across Champagne, and down the valley of the Marne. I enjoyed this flight immensely, though it nearly ended disastrously. The aeroplane I brought back was regulated for the weight of two men, so that when I flew in it alone I had to fight it all the way to keep it from climbing too far. Every moment I had to keep pushing against the control and it almost exhausted me. There was a low ceiling of clouds and I simply could not let the machine have its own way. To add to my aggravation, the motor stopped as I was passing over a forest. There was nothing to do but volplane down, though I did not see how I could ever avoid the trees.

Unexpectedly a clear landing-place loomed up ahead of me, but before reaching it I felt that I would be in the tree-tops. Worst of all, I had a lee shore. Across my path I suddenly noticed a canal lined with poplars. I could not possibly pass over them, so I pressed desperately against my rudder controls. Being near the ground, it was a frightfully short and sharp turn. I thought that the tip of one of my wings would touch the branches of the trees while the other would scrape the ground; then I would be crushed under the motor. At that moment the machine straightened itself out and came to a stop in a ploughed field. It was a very close call.

I shall never forget one of my flights over Paris. The day was beautiful. The atmosphere was so clear that one could see for miles and miles. As I approached the city it looked like a toy model. Every street, almost every house, stood out in perfect detail. The white church on the hill of Montmartre glistened like ivory. Beyond it I could see the *Arc de Triomphe* and the *Tour Eiffel*.

I stayed up so long that my supply of gasoline was almost exhausted, and I was obliged to land to refill the tank. I chose the aviation-field at Le Bourget, the scene of my first war experience on the night of the Zeppelin raid. As it happened, I again selected an unusual occasion for my visit. This time, however, the extraordinary activity was not due to an unwelcome visit by the Germans. It was rather to celebrate the perfection of an unpleasant surprise for the hated Boches.

Great crowds lined the field on every side. In the centre stood a small group of prominent officials. Among them I recognized President Poincaré. They were examining a new weapon with which French aeroplanes would henceforth go "sausage- hunting" over the German lines.

Even the casual visitor to the front is struck by the great number of observation balloons which both sides use in their efforts to keep informed of the preparations being made by the enemy. Every few miles a captive balloon or "sausage" wafts lazily over the German lines, fairly far behind the lines, but at an altitude sufficient for observation purposes. Against these "monsters" aeroplanes heretofore had been powerless. Their machine guns fired bullets which, even if incendiary, were too small to set on fire the gas-containing envelope. The aircraft cannon carried by some of the French machines also proved useless. The holes their projectiles made in the balloons were too small to allow a sufficient quantity of air to enter and cause an inflammable mixture.

The rockets, which were being examined as I landed at Le Bourget, solved the problem. Four are mounted on each side of an aeroplane. At the head of each rocket is a large dart, resembling a salmon-gaff. The tails of the rockets are wound into spiral springs, which are held in sockets. All eight rockets are fired at once. They are ignited as they leave their sockets, and travel with lightning speed.

Swinging lazily above the field was a captive balloon. At one end of Le Bourget was a line of waiting aeroplanes. "This is the second. They have already brought down one balloon," remarked the man at my elbow. The hum of a motor caused me to look up. A wide-winged double-motor Caudron had left the ground and was mounting gracefully above us. Up and up it went, describing a great circle, until it faced the balloon. Every one caught his breath. The Caudron was rushing straight at the balloon, diving for the attack.

"Now!" cried the crowd. There was a loud crack, a flash, and eight long rockets darted forth, leaving behind a fiery trail. The aviator's aim, however, was wide, and, to the disappointment of everyone, the darts fell harmlessly to the ground.

Another motor roared far down the field, and a tiny "*appareil de chasse*" shot upward like a swallow. "A Nieuport," shouted the crowd with one voice. Eager to atone for his "*copain's*" failure, and impatient at his delay in getting out of the way, the tiny biplane tossed and tumbled about in the air like a clown in the circus-ring.

"Look! He's looping! He falls! He slips! No, he rights again!" cried a hundred voices as the skilful pilot kept our nerves on edge.

Suddenly he darted into position and for a second hovered uncertainly. Then, with a dive like that of a dragon-fly, he rushed down to the attack. Again a sheet of flame and a shower of

A HEAVY BOMBARDING MACHINE

sparks. This time the balloon sagged. The flames crept slowly around its silken envelope. "*Touché!*" cried the multitude. Then the balloon burst and fell to the ground, a mass of flames. High above, the little Nieuport saucily continued its pranks, as though contemptuous of such easy prey.

To the north a group of tiny specks in the sky seemed to grow in size and number. Nearer and nearer they came. I knew they must be a bombarding *escadrille*, returning from a raid across the enemy's lines. One, two, three, I counted them, up to twelve. Slowly they floated along as if tired by their long flight, and then gently they began to drop down.

They rolled smoothly across the field and stopped before their hangars. Cannon protruded menacingly from their armoured, boat-shaped bodies as the pilots climbed down and stretched themselves.

"At the front this morning, tonight they can dine in Paris," jealously sighed an infantry officer.

"But," replied an aviator from our group, "there are two of them who will probably never dine in Paris again. Fourteen started out this morning. Now they number only twelve."

Ordered to the Front

After three weeks at Plessis-le-Belleville I became "*disponible*," that is to say, I was listed among the first twenty aviators who were considered ready for duty at the front. From that moment orders directing my future movements might be received any minute, and I was under restriction not to stray too far away. I must say that I experienced a curious sensation, waiting around in this way, not knowing where I would be in a week. You never know to what sector of the front you are going until your orders are handed to you. Three days after my name had been posted on the bulletin-board an order came detaching five pilots for duty with the "*Armée de l'Orient*" at Salonica. My name was sixth on the list, so I missed by one being among them. That evening, however, my turn came. This time the direction was Toul.

When men leave the *Réserve* for the front there are no sad leave-takings. Every pilot seems to be glad that his turn has come to do his share in the defence of his country, and instead of being downcast he is light-hearted. Yet the part which he is to play in the air involves chances which are four to one against his coming through alive.

It is customary for a pilot to spend two days in Paris before starting for his ordered destination. Officially my leave was due to begin only on the following morning. I decided, however, to take time by the forelock and to be off that night. In this way I would gain twelve hours additional leave. All the "*paperasserie*"— red tape—was first disposed of, and then I proceeded to pack my effects. These I had drawn from the aviation quartermaster's depot. There was my fur-lined union suit, a fur overcoat, fur boots, gloves, and cap. I also received an automatic pistol with a holster, a special aeroplane compass, an *"altimetre,"* a special aviation clock mounted on wire springs, and a speed-indicator. These were furnished to me by the government, and became my property. I had the privilege of providing myself with anything else I wished, but the government outfit always had to be at hand for inspection.

Fortunately I had just time to make the evening train for Paris. According to my pass, I was on *"service recommandé"* and on my way to the front, where in a few days I would be flying over the German lines. The moment I had looked forward to for so many months had at last come. I could hardly believe it myself.

My two days in Paris passed like magic. There was so much to attend to, so much to do, that before I knew it the moment to leave had come. I took a taxi to the *Gare de l'Est*, my wife and my little girl accompanying me. My luggage consisted of my black army canteen, across the front of which was painted in white letters "Carroll Winslow—*Pilote Aviateur,*" and my long canvas duffel-bag, which contained my fur-lined clothes and all my flying paraphernalia. There are usually so many formalities to be complied with that I allowed more than enough time for the visa of my papers.

It was well that I did. The station was crowded with grimy, blue-coated *"poilus,"* walking up and down the waiting-rooms and lounging on the stone steps; outside others were saying goodbye to their families, while across the street large numbers crowded about the free "buffets," where patriotic women of Paris daily minister to the wants of the departing *"permissionnaires."* All the men wore their steel helmets, and had their knapsacks strapped to their shoulders. They were not as smart-looking as the khaki-clad British "Tommies," but despite their muddy boots and faded uniforms there was something in their faces, a look in their eyes, that seemed to say: "No sacrifice is too great—for France." I felt proud to think that I was one of them, and their quiet salutes showed me that I had their respect. The regard of those grave, war-worn men meant much to me.

My wife and I silently watched what was going on about us, while our little girl chattered at our side. Many women accompanied their "braves" to the station. Most of them carried baskets of food and delicacies, but some, too poor even to buy a present for their "*poilus*," came empty-handed. The moments of leave-taking seemed almost tragic. Many a man went up those steps whistling and with head erect, while others laughed as they tossed their little ones high in the air for a last goodbye. These were fine examples, and when the porter touched me on the arm and said: "The train for Toul, *m'sieu'*," I too was able to bear it calmly.

The cars were already crowded with "*poilus*." Not a seat was to be had in the compartments. Standing-room in the corridors was at a premium. We were all bound in the same general direction, toward Verdun, Nancy, and Toul.

The train came to a stop at last. We were at Bar-le-Duc, the terminus for Verdun. What an air of mystery there was about the station at "Bar." We could hear the distant roar of the cannon defending the banks of the Meuse. Everywhere men moved about with a sort of suppressed excitement. "*Camions*" rumbled by in hundreds. In the freight-yard troops filled every available space not already taken up by the newly arrived artillery. Nearly all my travelling companions left me here. For a moment I wished that I too had been ordered to the Verdun sector. It was after sundown when the train drew into the station at Toul. The town was in darkness, and I felt very doubtful as to whether I would be able to join my *escadrille* that night. To my surprise, an officer, noticing my indecision, came up to me and asked if he could help me. I told him where I wanted to go and inquired if he could direct me. "Why, it is too late to do this today," he remarked; "better wait until the morning." With that he motioned me to step into his automobile and directed the chauffeur to drive us to the *Etat-Major*.

As we rolled through the streets of the silent city I had a moment to reflect upon what all this meant. I began to realize

A German aeroplane brought down by a French aeroplane.

The smoke is from the German machine, which the aviator has set fire to upon being brought down. The French machine can be seen to the right, its wing broken by a bad landing. The small dots in the centre are French soldiers. The white lines are the French third-line trenches.

The French pilot with his German prisoner (insert).

that a change had taken place in my position. I was no longer a mere soldier, but an aviator and as such entitled to courtesies usually extended only to officers in the other branches of the army. There is no mention of this custom in the regulations. It was merely an unwritten paragraph of military etiquette. Here was an officer, my superior in rank, treating me with a consideration I had rarely experienced. I noticed by the insignia on his overcoat that he was a captain in the Aviation Corps. He was therefore a pilot. I thought for a few moments.

Suddenly an idea occurred to me. I was also a pilot, and in the eyes of traditional convention we were comrades, for we were both aviators. At the *Etat-Major* the colonel likewise extended a warm welcome and shook me heartily by the hand. I suppose that my being an American had something to do with it, but I could not help thinking that I was still only a corporal. He immediately gave orders to requisition a large room for me at the hotel, and bade me hurry or I would be late for dinner. No wonder aviators are inspired to do such splendid work at the front when their efforts meet with so much appreciation.

The next morning I started out soon after sunrise to walk out to the aviation-field. Everywhere, above the streets of Toul, there were posters which read "*Cave Voutée*," and with the number of persons, varying from fifteen to sixty, who could be accommodated. These cellars were protected with sand-bags and were located at convenient intervals, so that the people might find shelter quickly whenever the German aeroplanes made their appearance. Only a few days previous to my arrival there had been a raid, yet everything seemed normal and the housewives went about their marketing and shopping as if they had nothing else to think about.

An hour's walk brought me to my *escadrille*, F-44. I was barely in time. Orders had just been received transferring it to the Verdun sector and preparations to move camp were already under way. Everyone went to work with a will, laughing and jumping around in a sort of war-dance. No wonder they were happy. They—we, I should say, for I was now one of them—were about to become participants in the world's greatest defensive battle.

The aeroplanes started for Bar-le-Duc "by air" shortly after noon. One pilot and myself, however, had to make the journey by rail. My own machine had not yet arrived, and his had been smashed up the day before. When the German raiders came over Toul he had gone up with the defending aeroplanes, and had brought down an *aviatik* which he had engaged. It is customary for a pilot, when driven down in the enemy's territory to set fire to his machine to prevent it from falling into the hands of his adversaries. This the German proceeded to do the moment he touched ground. My friend was frantic to prevent this and tried to make a quick landing in order to get to him in time. He was too excited, however, and smashed one of the wings of his own machine during the landing. This occurred just behind the French third-line trenches. The soldiers rushed out and made the German pilot a prisoner, but not until after he had applied the match to his gasoline-tank.

In The Verdun Sector

At Bar-le-Duc I felt again the suppressed excitement of the near-front. Everywhere were "*Cave Voutée*" signs, troops were in motion on all sides, sentries were posted at every street-corner, everyone seemed to be in a hurry to get somewhere.

Our *escadrille* was camped in a field adjoining that occupied by the American Escadrille. Our "train" consisted of a dozen light, covered trucks with their tent-like trailers, and three automobiles for the use of the officers and pilots. Our camp was pitched by the time I had made the trip from Toul by rail, and the array of tents and the park of tractors had every outward appearance of a country circus.

It was my first impression of an air-squadron camp at the front, and I must admit that my previous conception of the amount of equipment required by each of these units was far below what I now beheld. The personnel of my *escadrille* alone looked like an expeditionary force for service in Mexico. There were a dozen artillery-observers, seven pilots, countless mechanics, chauffeurs, orderlies, servants, wireless operators, photographers, and other "*attachés*," over a hundred and twenty-five men in all. Each of these hundred-odd men were essential to the work of the nineteen pilots and observers.

It was a pleasant surprise to find the American pilots here. I had not heard that they had been ordered to the Verdun sector. This honour had been thrust upon them unexpectedly. They were now here, among the best fighting units of the French Army, to protect the photography, fire-control, and bombarding-machines of this sector. Their camp was thirty miles behind the lines, but with their fast little Nieuports it took them less than fifteen minutes to be in the thick of the fray. The government had given them a large, comfortable villa to live in. I must say I felt a bit envious when I compared their feather-beds and baths with my little tent and canvas-covered cot.

That evening I had dinner with my compatriots. It was a meal I will never forget. As visiting pilot I was seated on the right of their commander, Captain Thenault. Across the table, opposite me, sat Victor Chapman, Norman Prince, and Kiffen Rockwell—all three since fallen on the *"champ d'honneur."* At the other end of the table were Elliot Cowdin, Jim McConnell, and "Red" Rumsey, together with Clyde Balsley, Chouteau Johnston, and Dudley Hill. Bill Thaw was not with us, as he was in the hospital, having been wounded in a recent combat with a Boche.

The places of the three pilots killed have since been taken by other volunteers, but in the minds and memories of the Americans dining at the camp that night their places can never be filled. We know that they did not die in vain, and that what they did will live in history. Their spirit was one of sincere patriotism to the cause they had made their own, and among the Allies the sympathy and the belief they expressed has been amply proved.

The *escadrille* was to make its first sortie as a unit in the morning. Captain Thenault had much to say to his men, and after dinner the conversation continued along the same general lines. There seemed to be so much detail to attend to and signals to arrange that I was almost tempted to ask them how *escadrilles* ever managed to co-operate so well in the presence of the enemy air squadrons.

When I awoke next morning it was raining. The clouds hung low, too low for flying over the lines, so the Americans remained in their beds. Our *escadrille*, however, was obliged to move on, as the station to which it had been assigned was directly behind the lines. The planes had to proceed *"par la voie de l'air,"* but the ground was so soft and muddy that it was difficult to get the

A bi motor Caudron

A captured Fokker

machines to leave the earth. The pilots all seemed nervous, yet all rose in good form except one, who was a little late in getting off. He did not know the way, and was afraid of losing his companions in the mist. In his haste he took too short a run, so that when he came to the end of the field he was not high enough to clear the line of hangars in his path. To make matters worse the unlucky man lost his head. He tried to make a sharp turn, but it was too late. The tip of his wing caught the canvas of the tent, and the machine fell with a crash to the ground, killing the pilot and pinning his mechanic beneath the wreckage.

We felt much depressed by this accident. Our departure for the new camp seemed to emphasize our sadness, for, as we moved off in our long line of motors our procession had an appearance almost funereal. First came the automobiles; then, following them, the twelve tractors and trailers—twenty-seven vehicles in all—moving slowly toward the front.

As we turned into the main road to Verdun the traffic was so heavy that we had to move at a snail's pace. Ahead of us rumbled a steady stream of "*camions*" with ammunition and supplies. Alongside of the road were the columns of troops going to the trenches. Their heavy coats were already soaked, and the probability was that they would remain so for a week, but nothing daunted them. They just plodded along gayly, singing their marching songs, utterly unmindful of the raindrops that were hourly weighting down their equipment more and more.

From the opposite direction came the empty supply-trains. Sandwiched in with these were ambulances and motor-buses, bearing the men returning from their "stage" in the trenches. The poor fellows looked hardly human, for they were brown with mud from head to foot. Their faces were caked with dirt, and a week's growth of beard gave them a still more uninviting appearance. They seemed to gaze at us with a far-away, half-conscious expression, so utterly stupefied were they by the terrible bombardment to which they had been subjected.

The farther we went the more numerous were the evidences of war. The roar of the cannonade became louder. On both sides of the roads the villages were in ruins.

Not a farmhouse was inhabited, and the fields were dotted everywhere with soldiers' graves; on each cross hung the "*kepi*" of the dead hero. In some of the military cemeteries there were graves without little wooden crosses—only a small fence marked them off from the rest. These, I was told, were the graves of the Mohammedan African troops, whose comrades claimed for them a plot apart from the "unsacred ground" used by their Christian allies.

It was almost dark by the time we reached our new camping-site. The fields were soaked with the heavy rain, and we splashed about in the mud for hours before the task of pitching camp was completed. By nine o'clock, however, all was ready and we sat down to a good, warm supper. Then we turned in. It was so cold and chilly that I went to bed in my fur-lined clothes. But tired as I was I could not get to sleep. The roar of the artillery was frightful. On every side of us it crashed and thundered, unceasingly, uninterruptedly. An attack was in process at the *Mort-Homme,* and every little while there would be a "*tir de barrage,*" or curtain fire as we call it. The small 75's would sound like the *rat-tat* of a snare-drum accompanying the louder beats of the deep-bass drums.

I got out of bed and gazed toward the battle-field. The earth was brilliantly illuminated by the rockets and flares that were being sent up everywhere. The sky seemed full of fire-flies—in reality exploding shells. On all sides the guns flashed angrily. Searchlights played about in every direction. It was a most superb spectacle, but it was terrible. It was hell.

My First Flight Over the Lines

Unfavourable weather conditions kept us inactive for several days, but as soon as the skies cleared our *escadrille* immediately went to work again. For some reason my own machine was delayed "*en route*," and did not arrive for a week. This was time I could ill afford to lose, so the "*chef pilote*" took me as a passenger in his biplane to familiarize me with the ground in our sector.

We started late one afternoon. The atmosphere was extraordinarily clear. Every detail in the landscape stood out boldly, and as we rose the dozens of camps in the immediate vicinity spread out below us like models set in a painted scenery. The valleys, the tents, the guns, the troops, all were visible to the naked eye. On all sides were aviation-camps, which were easily distinguished from the others—there must have been at least twenty of them within a radius of five miles.

As soon as we reached a height of three thousand feet my pilot headed the machine toward the lines. At our feet lay the terrain of the "Verdun sector." From the forest of the Argonne on our left to the plains of the Woëvre on our right stretched one of the bloodiest battle-fields of history. At regular intervals along the front the French captive balloons—there were eighteen in sight at this moment—swung lazily in the breeze. They looked for all the world like the "*saucisses*" they are named after. Day and night they are kept aloft, maintaining ceaseless vigil over the movements of the enemy.

Passing the balloons, we could see the various important points of the defence at closer range. The city of Verdun nestled close to the banks of the Meuse, which wound like a silver band through that now desolate land. Far off to the right were the forts of Vaux and Douaumont. A trifle nearer was Fleury. To the left, in the distance, I could make out the *"Mort-Homme"* and Hill 304, while directly before us lay Cumières and Chattancourt. The entire Verdun sector was spread out like a relief-map.

The German attacks upon the French position on the *Mort-Homme* were still in progress. I had never before seen a battle, and to see such an important conflict from "the gallery" seemed most strange. It looked more like a pan of boiling water, with the steam hanging in a pall over it, than anything else I can think of. In fact, a yellow mist rose to a great height and almost obscured the view. Tiny flashes showed where the guns were concealed, but to us the battle was a silent one. The noise of our motor drowned the whistling of the shells and the roar of the bombardment. I could not help thinking how much some of those poor fellows below us would appreciate a little of this silence.

We could plainly see the network of the trenches, broken and half-obliterated in the mud. In some places they were so close together that it was difficult to make out where the French lines ended and the German earthworks began. The ground was speckled with "pock-marks" caused by shell explosions, and altogether it was a weird scene of desolation. All signs of nature which had once beautified this region had vanished. The forests and the green fields had disappeared. Ruined villages lay like piles of disused stone among the circular *"entonnoirs,"* or shell-holes. In colour it was all a dirty brown.

On every side of us were the French artillery biplanes. They were hovering over the German lines like gulls, continually wirelessing back the ranges to their batteries. High above us circled the little Nieuports on guard, to protect us and to prevent the Fokkers and *aviatiks* from crossing over our lines. Everywhere were little white puffs, which seemed to follow the machines about. I watched them, strangely fascinated and amused, until my pilot informed me that these were caused by exploding shrapnel

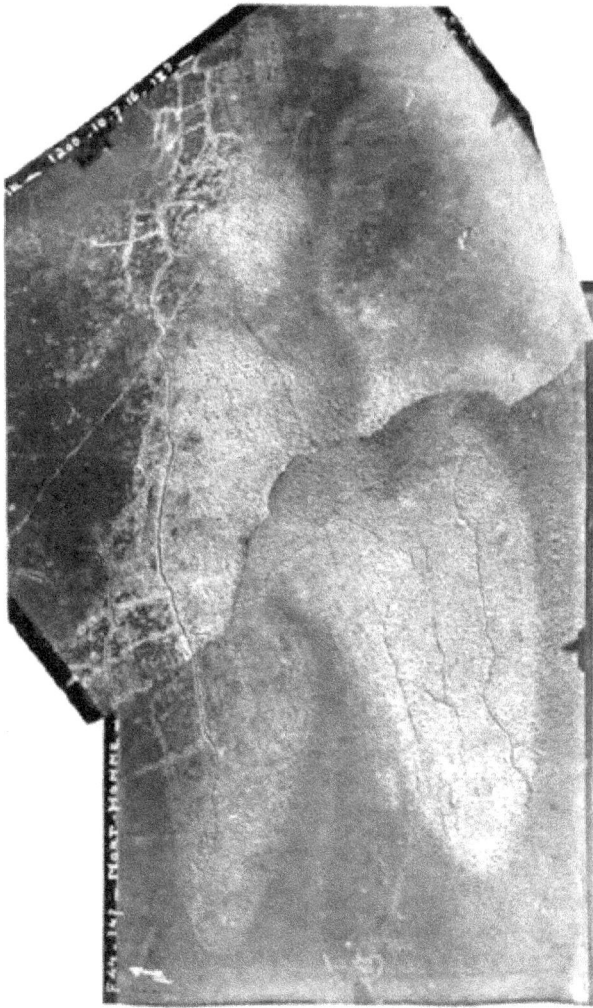

A view of the Mort-Homme taken from a height of 3,600 feet.

These are two photographs pasted together. Exact maps of the front are made in this manner daily by the photographic sections.

from the enemy's anti-aircraft guns. Then I noticed with uneasiness that the same puffs were also following us. My interest in the little white puffs from that moment assumed quite another character. I listened for the sharp crack of their explosions, but all I could hear was a dull "*whung.*" The thought that very few machines are really brought down by shrapnel was a bit reassuring, but I must admit that when the enemy is sending them on all sides of you, you do not feel like giving much credence to what others may have told you.

Presently my attention was called to the lines of German captive balloons, which are moored some miles behind their first-line trenches. Several aeroplanes stood guard over them, and as we knew that they were armed and that we on this occasion were not we decided to turn back.

I made several of these trips over the different positions on our immediate front. By the time my own machine arrived I was thoroughly familiar with the sector and also with the main dangers to be encountered by aviators over the battle lines. The first precaution I learned was—always, when landing, to unhook the belt that held me in my seat. This is one of the most important things to remember at the front. The fields are not always in the best condition, and the slightest obstruction may cause an unexpected crash. If you are in an artillery pusher-machine when this happens you are invariably crushed under the motor, unless your belt is unfastened, when you are usually thrown clear.

Another danger, which I would never have thought of if an experienced pilot had not pointed it out, lies in the cables mooring the captive balloons. These are invisible to an approaching aviator and to collide with one means a fatal smash. When flying low enough to pass under the "*saucisses,*" aviators must watch out for these "*tethers.*" Nevertheless, you can always take advantage of one of their peculiarities. The cable always stretches to windward, and in a good breeze it stretches far. By keeping well to leeward you can always rest reasonably assured that you are on the safe side. Many aviators, however, have met with fatal accidents, through fouling these cables. I know of only one instance where the pilot did escape unhurt after striking the wire. It seems that the moment he saw what was going to happen he put his machine into a vertical bank, so that when the impact came he was turning about the cable. Then, strangely enough, by continuing his spiral he was able finally to disengage himself and escape.

Telephone and telegraph wires also are a certain menace to aviators. They form a regular network behind the lines, while on every aviation-field there are in addition wireless aerials to avoid. Many a returning pilot has forgotten them in his haste to get back to camp, and fouled them, to his regret. One pilot I knew met his fate in this way. He had been wounded by a shrapnel-ball over the German lines, and had managed to return to his own field. He was so weak from the loss of blood that in his anxiety to land quickly he forgot the aerials. His machine caught the wires and fell to the ground. Both the pilot's legs were broken in the fall and he died, not so much from his wound as from this unfortunate accident.

Still another risk is encountered when flying in the clouds. A cloud is dangerous at any time because there may be an enemy—or, in fact, any machine—in it. If you enter the mist you may be going head on into another aeroplane without having the slightest warning of its presence. Your own motor makes so much noise that you never under any circumstances hear that of another machine until too late. You are in consequence deprived of both your eyesight and your hearing. At the front the risk of meeting an enemy aeroplane under such circumstances can never be overlooked, for often fighting machines use the mist to cover their presence.

Shells also have to be carefully avoided, for, though destined for some far-away target below you, they sometimes in their flight destroy aeroplanes unintentionally. As I have already explained, we devoted much time to this subject at Châteauroux, learning the trajectories of the different calibres. Still, at the front, the theory is not so easily put into practice. It seems almost impossible to keep track of all the artillery massed by your own

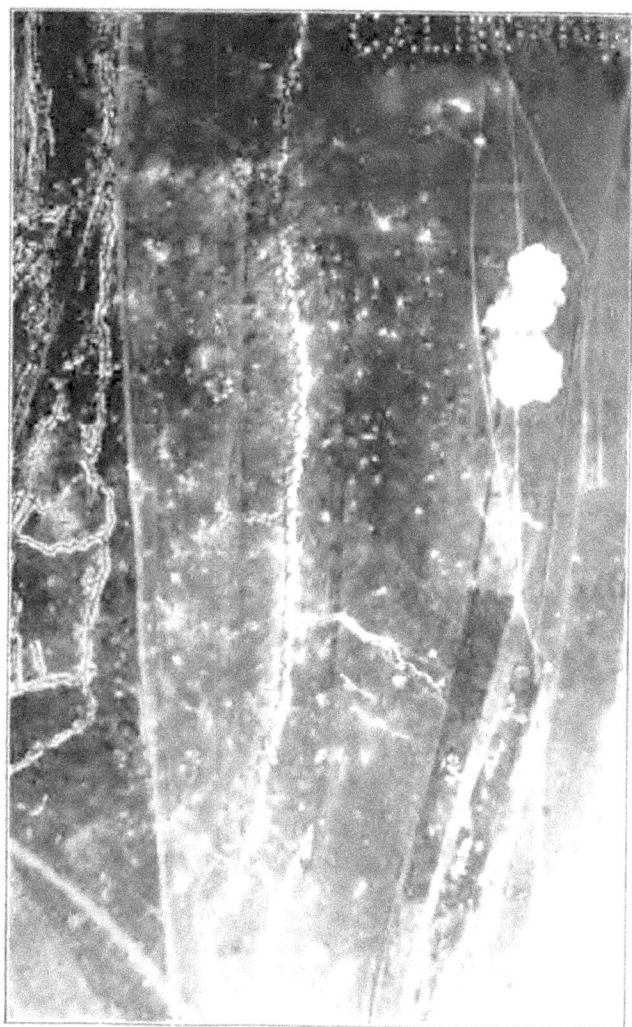

"EVERYWHERE LITTLE WHITE PUFFS SEEM TO FOLLOW THE MACHINES ABOUT."

side, especially in such a sector as Verdun, where the guns often were placed so close together that their wheels almost touched.

On more than one occasion when flying quietly through the air my machine has given a sudden lurch, and I have heard the dull "*tung*" of a passing shell. There is none of the whistling we are accustomed to on the earth; merely the dulled sound caused by the sudden displacement of the air. My own machine finally arrived, after delays that seemed interminable, and my two mechanics immediately set to work installing the various instruments, and painting it. These two men were personally responsible to me for the condition of the motor and planes, but, as pilot, I was the master of the machine, which was reserved for my own use. In fact, each aeroplane has painted on its body and rudder the name and distinguishing marks of its pilot and *escadrille*.

After a few short flights I became aware of the fact that my biplane, in spite of all my efforts to correct it, showed a strong tendency to lean to the right. At times I could hardly make a turn to the left. This was a serious matter at the front, as an enemy might at any moment appear on my "weak" side and I would be placed in a serious position. I therefore mentioned the matter to my captain.

To my surprise, he immediately ordered a new machine for me and gave directions that the one I was using should be sent back to the factory. The defect in this particular case was one mechanics could not remedy, and it seems that it was nothing out of the ordinary to send a machine back to the shops. At the front a pilot must have a perfect machine to work with or none at all. The life of a good aeroplane seldom is more than fifty hours of actual flying.

During this time the organization of the escadrille was perfected. The pilots were divided into two "watches," one-half being on duty while the other was "standing by" ready for service in case of emergency. All the pilots except myself were "*disponibles*." I was exempt because I had no machine, and was therefore for the time being my own master, even when it came to rising in the morning. When the others on duty were awakened, at early dawn, I would be awakened with the rest. My turn had not yet come, however, and I could just turn over and sleep to my heart's content.

Our camp looked like a little tented city; there were seven enormous canvas hangars, and grouped about these six other tents, each serving a particular purpose: captain's office, wireless plant, telephone central, repair-shop, photographic division, and kitchen. At one end of the field were the living-quarters of the captain and the observers, while at the other were parked the thirty automobiles of our two *escadrilles*. On the opposite sides of the field were the quarters of the pilots of the two *escadrilles*. The mechanics slept in the hangars with their machines.

Considering everything, we were fairly comfortable. The pilots of each *escadrille* shared two large tents, and in addition each group had a large mess-tent. Inside each sleeping-tent each one of us had a little alcove. Our cots were raised on wooden platforms. At one end we fitted up a shower-bath, for which purpose a gasoline tank punctured with holes proved ideal. Of course, every time you wanted a bath someone had to empty pails of water into the "tank" above you. Our mess—"*popotte*" they call it in the French army—was very good. We had a regular daily allowance from the government, but this was not always enough to buy all the supplies we needed. We therefore instituted a system of fines, and our treasurer provided our table with a small tin box in lieu of a centrepiece.

Bad language or talking "shop" before coffee involved a ten-*centime* fine, which had to be dropped into the bank at once. This regulation proved a godsend to the mess—and to our conversation.

As I was not "*disponible*," I was sent on several trips with the staff automobile. Its most frequent runs were to the artillery headquarters to deliver photographs of the enemy's positions. These were situated in a nearby village, within sight of the German trenches. All the roads approaching this place were masked, and the town itself was in ruins. Everywhere sand-bags reinforced the stone walls.

The telephone central was a veritable fortress, and continually within the zone of the German artillery "strafes." The life of the officers of the *Etat-Major* was certainly not an enviable or an easy one.

Co-operating With the Artillery

By the time my second machine arrived I had been at the front long enough to appreciate the *rôle* played by each of the different types of aeroplanes used in this great conflict. Camped near us was a bombarding unit. Every night when the heavens were clear these machines would go up, turning great circles over our heads until they reached the desired altitude. They would then vanish with their destructive bombs in the direction of the enemy. We could always tell when they passed over the lines, for the German searchlights would become very active, and the sky would become dotted with sparks, which in reality were exploding shrapnel. Then, in the early hours, of the morning, they would return, having flown far into the enemy's country to drop their bombs. This is tiresome and disagreeable duty, but not by any means as dangerous a one as the other branches of aviation, for bombers are practically free from interruption by enemy aeroplanes in the dark.

Camped with us was the famous N-64, the crack fighting unit of the French Aviation Corps. Among its pilots were such famous aviators as Navarre, Nungesser, and Vialet, known familiarly as the "aces." Every evening just before sunset these men would take their machines up in the twilight and do "stunts" for the benefit of hundreds of admiring *"poilus"* gathered from neighbouring camps. These were the self-same "stunts" which on many occasions had enabled them to escape from sure death at the hands of some superior enemy force.

As I have said before, the fighting work, although in reality the safest, requires the most experienced and accomplished pilots. The chief duty of the Nieuports is "barrage," or sentry duty. There are always several of them flying over the lines, on the lookout for some "Boche." It is their task to swoop down like a hawk upon them and destroy or else drive them away. Of course, our own "*avions de chasse*" are as liable to be attacked by the enemy, and they must in consequence be continually on the "*qui vive.*"

When one of the larger reconnoissance machines is compelled to go far into the German lines on special-mission work, it is usually accompanied by a body-guard of several Nieuports. Spies, on the other hand, are carried only in the fast-flying Nieuports, which in this case are double-seated. It seems that it is comparatively easy to take a man over and leave him far in the rear of the German trenches, but going back for him is another matter. After several days the pilot returns to a prearranged place, but, as sometimes happens, his compatriot may have been caught. In this case a like fate usually awaits him at the hands of the watchful enemy.

For reconnoissance work the large bi-motor Caudrons are generally used. They are fitted with a small wireless apparatus; but this means of communication cannot be used very often. The machines, on many occasions, have to go beyond the effective radius of their radio, and at other times its use is inadvisable, as its messages might become known, or else blocked by the enemy. Resort has been had, therefore, to carrier-pigeons. These are released the moment any important information has to be conveyed to headquarters, and these swift little messengers have proved extremely useful and reliable. Their use has, in consequence, become general.

The bi-motor Caudrons are employed also by the photography section of the army, though much of this work is actually done by artillery machines detailed for this service. Photography is a dangerous duty, because the flights have to be made at low altitudes to obtain the best results. On the other hand, it is not at all tedious. The mission upon which the machine has been sent is usually accomplished in a brief space of time, and the machine often stays out less than an hour. In comparison with the amount of flying required of the aviators in the other branches of the service, which varies in length from three to five hours a day, photography is easy work.

"*Réglage*," or fire-control, on the other hand, is the most difficult and the most dangerous work to be performed by the Flying Corps at the front. The machines used are large and unwieldy, built to carry the weight of two men and all sorts of equipment. They are fairly fast, but their spread of wing is so large that it is almost impossible for them to make a turn quickly when attacked. They are armed with a machine gun, it is true, but they are always at a great disadvantage in the presence of an enemy fighting-machine which can outmanoeuvre them at every turn.

The first duty to which I was assigned was "*réglage*," and this, I found, involves many complications. The chief source of trouble usually is the wireless apparatus, which has to be maintained in perfect working order. Before leaving the home field you usually circle over it, while your observer tests his sending apparatus. The receiving operator then answers by visual signals. Usually these are large white sheets laid on the ground in different formations, which have a prearranged meaning. When the radio is found to be in perfect order you are off to the battery you have been ordered to co-operate with. By wireless your observer then reports to the battery commander, and receives his orders by means of the same visual signals. You then head in the direction indicated to you before leaving, and, hovering over the position to be bombarded, the observer signals back "fire."

The moment the shells have landed you turn quickly about and inform the artillery just how many metres their fire was long, short, or to the right or left. Your message is once more answered with the sheets. Again you fly back toward the enemy's position, circling in this way backward and forward between the battery and the target until the "*réglage*" is completed. Naturally every care must be taken not to disclose the position of your own guns to the enemy, or retaliation—"strafe," the English call it—summarily follows. Sometimes it is the battery which interrupts the work with the signal, "*Avion ennemi*," when the fire instantly ceases until the German aeroplane has disappeared or been driven off.

With such occasional interruptions the work continues until your observer can send back the signal "fire correct," which is generally answered by the "sheet signal" with the information that you may return home. Until this dismissal occurs, however, the ground below wholly engrosses the attention of your observer. You yourself are forced to keep a close watch for Boche fighting-machines so as not to be caught unawares by one of them. This is often a very trying task, as the models of some of the French and German aeroplanes are so very much alike that they cannot be distinguished until they are within range. The tricolour cockade and the black iron cross painted on the top and bottom of each wing serve to identify the fliers of the two belligerents, but these colours cannot be seen very far. You consequently have little warning as to whether the approaching planes are friends or foes.

Sometimes the enemy's anti-aircraft batteries become a bit too familiar. On such occasions the observer tries to signal to his batteries to drop a shell or two where these pieces are mounted. Often quiet is not restored until the machine has been more or less riddled with shrapnel bullets.

One "*réglage*" is very much like another, and when you have read the description of one you become familiar with them all. It is only in the results accomplished that the details vary.

It is a curious fact that in the first months of the war many artillery officers refused to follow the directions of their aerial observers. A colonel of artillery who has been firing big guns all his life cannot be blamed for not thinking that a young observation officer and a mere aviator know enough about the work of batteries to tell him where his shells are falling. Orders, consequently, had to be issued placing the artillery absolutely under the direction of the observers and calling upon the pilots to report any case where a battery refused to be guided by the signals it received. That put an end to the trouble.

At first I felt a strong aversion to flying over batteries in action. You are bound to get in close proximity to the trajectory of the shells, and the constant sensation and sound of the passing projectiles is none too pleasant. You get them both coming and going, and, no matter which you are trying to avoid, you are always taking a chance with the other. It is a question of choosing between the devil and the deep sea, with the devil constantly stepping into your path.

When you are observing for the artillery you must stay and die, if necessary, in the performance of your assigned duty. It is another matter with reconnoissance or photographic work. Here the main thing is to get back to headquarters with the information you have gained. If you are attacked and you see no chance of successfully fighting off the enemy, it is your business to run.

After some weeks of service with the fire-control detail I was ordered to serve as a photography pilot. This I found a most interesting duty. Whenever we received orders to photograph a position we would start out immediately, flying very low—say, from one thousand eight hundred to three thousand feet. As we reached the part of the enemy's positions to be photographed I would fly in parallel lines, while my observer took the photographs with a specially constructed telephoto camera. We would then hasten back to camp and immediately hand the plates over to the sergeant in charge of the dark room. This taciturn non-com would waste no time with words. In a few moments the photographs would be ready and on their way to headquarters. On several occasions I have seen photographs placed in the hands of the *Etat-Major* within an hour and a half after the order had been issued by the commanding officer there—examples of celerity and efficiency of service which have placed the photographic branch of aviation "*hors concours.*"

ESCADRILLE ═ M. 44

CUMIÈRES

LE 1 MAI 1916.

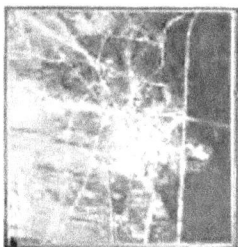

LE 30 MAI 1916.

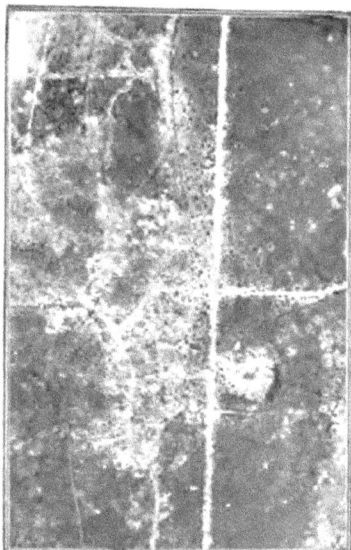

LE 20 JUIN 1916

REDUCED FACSIMILE OF THE PHOTOGRAPHIC REPORT SUPPLIED TO THE HEADQUARTER STAFF OF THE FIGHTING AT CUMIÈRES.

All in the Day's Work

It is strange how easily you become accustomed to being at the front. At first you sense your proximity to the vast military operations that are in progress, but after a while the newness wears off. One day passes like another without special notice, although daily something out of the ordinary is occurring somewhere along the western front. These experiences, however, generally fall to the lot of the fighting-machines. We of the artillery and photography sections share only the dangers. It is all in the day's work.

I remember one curious incident that occurred while I was in the Verdun sector. Victor Chapman, who was doing combat work with the American Escadrille after a brush with four German aeroplanes, was forced to descend to our field. Not only had he received a bad scalp-wound from a bullet but his machine had been riddled and nearly wrecked. One bullet had even severed a metal stability control. By all the rules of aviation he should have lost control of his aeroplane and met with a fatal accident.

But Chapman was an expert pilot. He simply held on to the broken rod with one hand, while with the other he steered his machine. This needed all the strength at his command, but he had the power and the skill necessary to bring him safely to earth. A surgeon immediately dressed his wound, our mechanics repaired his machine. The repairs completed, he was off and up again in pursuit of some more Boches. I must say that every one considered him a remarkable pilot. He was absolutely fearless, and always willing and able to fly more than was ever required of him. His machine was a sieve of patched-up bullet-holes.

A few days later came his last fight. He was carrying two bags of oranges to Clyde Balsley, who lay wounded in a hospital not far away. There was an aerial combat against odds within the German lines, and Chapman lost no time in going to the aid of his hard-pressed comrades. He brought down one of the enemy airmen, but the others were still too numerous, and the fight then was only a matter of seconds. He was last seen falling behind the German lines.

Balsley had been wounded in an encounter with several Germans. He was doing well, when he was struck in the thigh by an explosive bullet which burst in his stomach. He immediately lost all consciousness. His machine began to tumble straight toward the lines. Just before reaching the ground, however, Balsley regained his senses sufficiently to realize what was happening. By a superhuman effort he managed to right his machine and make a landing in a neighbouring meadow. He was carried to a nearby hospital, where for days he wavered between life and death. Two fragments of the explosive bullet were removed from his intestines. These he kept wrapped up in a handkerchief as proof that the enemy, despite their denials, do, violate the rules of civilized warfare. For a long time the only nourishment he could take was the juice of oranges, and that was why Chapman was on this mission on that unfortunate day.

A sad accident occurred on a neighbouring aviation-field while I was at the front. The captain of one of the *escadrilles* had visiting him his younger brother, a bright lad of nineteen. The boy was unusually well-informed about aeronautical matters, but he had never made a flight. His request to go up was acceded to, but the captain did not want to take him, so he asked one of his officers, the best pilot in the *escadrille*, to take him as a passenger. I suppose that the lieutenant was on his mettle, for before his machine was three hundred feet from the ground he began to do stunts. He was a past master in his art, but a bit too bold. Suddenly his machine slipped off on the wing, and crashed

HANDBILL DROPPED IN GERMANY BY FRENCH AVIATORS

to the ground. Even the best pilot was not immune against fate.

Our *escadrille* also met with a heartbreaking tragedy. One of our pilots, who had only recently joined us, was making his first flight over the lines with a young artillery officer, who was also inexperienced. Unluckily they flew too low and were brought down by rifle-fire. No one yet knows whether the pilot was mortally wounded or if it was the machine that was disabled. At any rate, the aeroplane came down in no man's land, between the French and German lines. The *poilus* immediately made a sally to rescue the two men and save their maps and important papers. The Germans had like intentions and opened a murderous fire upon them with their machine guns, trying themselves to reach the aeroplane. The result was a hand-to-hand struggle, and then a deadlock. Each feared that the other would reach the goal under cover of darkness. For a while there was a lull on both sides.

Then an inferno burst loose. Machine guns and field-pieces showered the unfortunate aviators with shell and shrapnel. In a short while the machine and its occupants were completely annihilated. The men, I believe, were alive when they landed, but it was impossible to save them. If the pilot could have steered fifty yards to the right or left, they would have been inside either line and their lives would have been spared. As it was, there never will be a monument to mark the spot where they perished at the hands of both friend and foe.

Occasionally the bombarding *escadrilles* have thrilling experiences to narrate. I remember one case in particular. The raiders were returning from a long flight into the enemy's territory when they were attacked by a group of German fighting-planes. An incendiary bullet pierced the gasoline-tank of one of the French machines and ignited it. The pilot knew that he was sure to be "*grillé*" and that he did not have time even to reach the ground. His minutes were numbered. Without a moment's hesitation he turned his machine sharply about and headed straight for one of his pursuers. The German tried to avoid the head-on collision, but he was too late. There was a sickening crash and both machines fell to earth.

Another case of desperate courage that attracted widespread comment occurred about the same time. This also related to a bomber who had been over the German trenches. The pilot was about to spiral down for the landing, when his passenger looked out to see if everything was in good order. To his horror, he noticed that two of the bombs were still unreleased, having become caught on the chassis or running-gear of the machine. If they landed in this condition, there was every likelihood that there would be nothing to mark their landing-place but a deep crater in the ground.

The two men were desperate. To climb down and unhook the bombs seemed impossible. No one had ever been known to do it. It was like clambering up to the main truck of a sailing vessel in the teeth of a hurricane. It was the only alternative left to them. The passenger mustered up his courage and climbed out on the wing and then down on the running-gear. Holding on with only one hand, he leaned down and carefully loosed the bombs with the other. It was a splendid exhibition of nerve and courage, and it saved the lives of both men.

Now and then you meet a pilot who has had a real adventure, but this is something only the most venturesome have to their credit. Not long ago, during an extensive reconnoissance behind the German lines, one of the pilots found himself flying parallel with an important railroad line. Presently he overtook a troop-train going in the same direction. Flying very low, he raked the cars with his machine gun until his magazine was empty. He then caught up with the engine and shot the engineer and fireman with his revolver. A little farther there was a sharp turn in the road, which the train took at full speed. Every car left the rails, and hundreds of soldiers perished when the train crashed down into the ravine below. The pilot confessed that he was sickened by the sight of the disaster, but it was war and he simply had to do it.

As far as my own experience at the front is concerned, it was unusually uneventful. My machine was never once hit by shrapnel nor was it attacked by the enemy. In fact, the work was very monotonous, one day being exactly like another. After six weeks I applied to my captain for permission to pass into a fighting *escadrille*, where the experience I had gained on the slower machines would be very useful and the work more agreeable. To my delight, my request was granted, and forty-eight hours later I received my orders to proceed without delay to the *Ecole de Combat* at Pau for further training.

It seemed rather strange, after weeks of actual service, to be leaving the front to go again to school. I had become so used to the life that the muddy fields and the little tents began to seem like home to me. Now that it was over, the *"popotte"* served to us in the mess-tent was most palatable, and I knew that I would miss the restraining influence of our system of fines.

The captain took me to Bar-le-Duc in his own automobile. As we left the field of our activities I looked back at our little camp. The mechanics were busy in the great canvas hangars, cleaning and repairing the aeroplanes and motors. Others loitered outside waiting for the return of their "patron" or for their pilot to go up. No one complained of the work or of the danger. It was indeed a privilege to be with such men. I felt a pang of regret at leaving them. Though they called out a cheery *"Au revoir!"* and *"Bonne chance!"* I knew that the parting was not so light-hearted.

July 14th, 1916

My trip back to Paris was very much like the one I had made in the opposite direction about six weeks before. Bar-le-Duc seemed unchanged as far as the outward signs were concerned. The movement of troops was just as great as during the previous weeks, only this time the regiments were leaving Verdun. The German efforts to take the fortress had failed signally and the offensive had passed to the French in the region of the Somme.

My train was very late in starting. Although scheduled for five in the afternoon, it did not actually get off until after midnight. It was filled to overflowing with *permissionnaires* and the crowded cars reminded me of a New York City rush hour in the subway. Fortunately there was a dining-car attached to the train. As this was kept open all night, we did not have to go hungry, and every one kept in the best of humour. It was interesting to see how quickly the men forgot what they had been through at the front. Within a few hours the *permissionnaires* were thinking only of the holiday which they were going to enjoy, of the good times they were going to have on the *boulevards*, and of home. The horror of battle was entirely left behind.

When we arrived at the *Gare de l'Est* it was barely five o'clock. The *quais*, however, were crowded with women who had apparently waited all night to greet their loved ones. Everyone seemed so happy. The men made no attempt to control their feelings. Tears veiled many a pair of eyes. How strange the contrast between this return and the departure for the front that I had witnessed not very long before!

Before leaving the station I had to have my papers stamped by the military authorities. This done, I hurried to a hotel. I was so tired after the journey that I could hardly keep my eyes open. It was not long before I was fast asleep.

When I awoke it was already late. I dressed and went out on the streets. To my surprise, large crowds lined the sidewalks. All seemed so gay. This was almost too sudden a transition from the type of crowds I was used to seeing in the Verdun sector. Then I remembered. It was the fourteenth of July, the *"Fête nationale,"* always a great day for the French people, but especially so this year. Someone soon informed me that there was to be a great review of the Allied troops, and that everyone was in consequence *"en fête."* At the front, however, I had heard little of this.

At the *Place de la Concorde* the throng was immense. The more enterprising had provided themselves with boxes and ladders to stand and sit on. Others goodnaturedly climbed up on the lampposts. The rest craned their necks in an effort to miss nothing of what was going on.

Earlier in the day the statues of the cities of Strasbourg and Lille had been bedecked with flowers. At the *Petit Palais* the President of the Republic had decorated, as is now the custom, the wives and children of those who had fallen on the *"champ d'honneur"* before their gallantry and patriotism could be rewarded.

As I reached the place the head of the parade swung out from the Champs-Elysées. It was the most impressive spectacle I have ever witnessed. Everyone in the crowd showed his emotion. The women could not conceal their tears, and the men only with difficulty restrained their feelings. First came the Dragoons, followed by the Belgian Bicycle Corps. Then the khaki-clad French African troops, with only their red *fezes* to remind one of their once showy uniforms. Their *mitrailleuses* came next, brought back from the front to accompany the gallant regiment on this occasion. The crowd then commenced to roar.

A battery of 75's then came into view, the *"soixante-quinzes,"* which to the Frenchman symbolize victory. Suddenly the crowd became attentive and quiet. The Russians were singing their deep battle-hymn as they marched. They were fierce-looking giants, and as they swung by to the wild, measured beats of their chants, the people were silent with admiration.

After the "barbarians," as the Germans call them, followed the Anglo-Saxons, clad in their khaki uniforms, the perfection of utility and smartness. There were the English, the Australians, and the Canadians, and, following them, a regiment of Indian cavalry. Then came the Scotch, headed by their pipers. They marched perfectly, swinging their legs in unison. Each time their right feet came forward a hundred white tartans rose together and exposed to the view of the astonished populace a hundred kilts and a hundred bare knees.

After the Allies came the French. The first regiment was from the Twentieth Corps. These were the men who had saved Verdun. There were many other units represented, many other regiments and brigades, but none received the welcome and the enthusiasm caused by the appearance of "Pétain's Iron Brigade" as they marched by in their quick, business-like step, with bayonets fixed to their rifles.

There have been many parades in Paris during the past decade, but there never was one like this. It was not a review—it was a war. Yesterday all these men were at the front. Tomorrow they would be back there again. For them this was only a momentary drop of the curtain on the tragedy in which they had been called upon to be participants. I could not help thinking of these poor fellows, some of whom I had very likely seen before, passing me in ambulances and motor-buses, muddy from head to foot and benumbed by the shock of battle. How many of these that I was seeing today would be in the ranks at the next review? I doubt whether these thoughts were in their minds. Tomorrow they would be on their way to take part in the Battle of the Somme, but with refreshed spirits and light hearts.

There was very little gayety or colour in the parade. All the troops were in their service uniforms. This was an hour of heroism and suffering, an hour of fixed determination which impressed upon one the feeling that the Germans could never win the war.

After the troops had filed by I joined some friends on the boulevards. It seemed difficult to believe that only a few miles away the hostile lines were linked in a death-grapple. Paris seemed normal. Of course there was not the animation that we formerly associated with the French capital, but there was little to remind one of the great conflict—only the aeroplanes patrolling overhead and the hundreds of *permissionnaires* wandering about the streets in their weather-stained and battle-stained uniforms.

That night I had to leave Paris for Dijon, to report at headquarters before going to Pau. By a strange coincidence I arrived at the same hour as on my first appearance to enlist. Now I viewed everything with different eyes. Instead of being a mere *"petit bleu,"* as they call the young soldier, I was now a *"pilote"* who had been at the front, I felt privileged therefore to walk right up to the buffet, and before long I was sharing a bottle of wine with a captain, a sergeant, and a second-class *poilu*. Such is the democracy of war.

The Finishing Touches

Great changes had taken place at Pau since my first visit six months before. The school had been improved and enlarged. Permanent sheds had replaced the canvas hangars, and the German prisoners had built a narrow-gauge railway from the town out to the field. The trains ran out to the aviation school every morning and afternoon, and returned before luncheon and again in the evening. This was a great convenience for many of us and in bad weather saved us many a long, weary walk. When the days were clear, however, we often made the journey on foot, as in this way we had sufficient exercise to keep us in good physical condition.

Only men who had already qualified as pilots or who had had previous experience at the front were allowed at the *Ecole de Combat.* We enjoyed practically the same liberties as at the front. We were free to go where we pleased, except during the working hours, when strict attendance and discipline were enforced.

I thought that my previous experience with the heavier machines would enable me to omit some of the more elementary courses, but this was not the case. I had to start at the very bottom. The management of a monoplane or of a small Nieuport is more delicate than anything I had ever tried, and the pilots have, therefore, to acquire a new "sense of touch" which is not required when flying in the larger biplanes.

My first assignment was to a Penguin, so called because it is nothing more than a Blériot monoplane with its wings cut down so that it cannot fly. The Penguins, however, are just as

A Penguin

difficult to manage as a full-fledged flying-machine, for on the ground your movements have to be more rough than when flying in the air. There are so many irregularities and air currents to affect your course that you have to be very quick with the controls. The Penguin, besides, does not answer the rudder as easily as the other types. I found that it was very difficult to keep a straight course when tearing across a field at the rate of about forty miles an hour. It was comical to see how the clumsy contraption behaved, turning circles, making "*chevaux de bois,*" rolling over on its wing, and behaving in every way like a drunken sailor trying to walk on a chalk line. You have to keep your head all the time, because the slightest misjudgement may result in an accident. When engaging a "*chevaux de bois,*" you must turn off your motor instantly, for neglect to do so will probably cause your machine to fall over sideways on its wing. When moving you must constantly keep the tail of the Penguin in the air in an imaginary line of flight, and if the tail is lifted too high you run the risk of sticking the nose of the machine into the ground and turning an unpleasant somersault. It was really interesting to discover how much skill it takes to manage a Penguin. It was several days before I could make the six straight lines required before you are allowed to pass into the next higher class.

In the second course a thirty-horse-power Blériot is used. I was made to fly in straight lines at very feeble altitudes, varying from twenty-five to fifty feet. The object of this instruction, it seems, is to teach the aviator how to take small, fast machines off the ground and bring them down properly. These smaller machines are able to climb much faster than the larger artillery types. This advantage is counterbalanced by the fact that they volplane much less, and are much more prone to slip off the wing. You have to handle them with the utmost care and gentleness. This point is very much emphasized in the instruction which you receive when flying in the 30-Blériots. Their motors are so small that you have to be very careful with them. You have to go about everything very gradually, except when making a landing. Then you must dive, and dive quickly, in order to retain your momentum.

As soon as I had been pronounced "apt" on the "*ligne droite*," I was assigned to the 50-Blériot. This, to my joy, included real flying. The difference between this machine and the ones I had flown in at the front was astonishing. There was practically no effort required of the pilot. The slightest move on the controls produced an instant response in the aeroplane. As in the case of the 30-Blériot, I found that the moment the motor was shut off, on account of the lack of volplaning qualities, to descend I had to point the machine straight at the ground. With the Farman I used to glide from unbelievable distances, but now I had to change my tactics completely and learn everything over again.

This course completed, I was granted leave of absence to return to America. Needless to say, I did not lose a moment in gathering my effects and engaging my passage. Next month, upon my return to Pau, however, I will have to take up my work where I left off. The first test required is a series of figure eights in a 50-Blériot and a number of difficult landings after this performance. Then follows a course in a Morane-Parasol. This machine, as I stated in an earlier chapter, is by far the most tricky machine in use today, (as at time of first publication). After you have learned to handle a Parasol, everything else is child's play. That is the reason why every pilot of a fighting *escadrille* is made to master them. It is the best experience to give you a sense of balance yet discovered.

Before you are allowed to fly in a Nieuport and attend the School Aerial Acrobatics there is another requirement. This is a brief period of instruction at the Mitrailleuses School at Casso, where, on the shores of the long lake, the French army has established an ideal range for the training of its pointers. It is less than an hour by rail from Bordeaux and well within the reach of every military depot in the south-western part of France. Each branch of the service has its own course. For the Flying Corps the range consists of a number of captive balloons and of a series of moving targets on the lake. The pupil is taken up as a passenger in a double-seated aeroplane and operates the *mitrail-*

leuse. After two or three weeks of this practice he becomes quite used to shooting from an aeroplane and finds that he can score hits almost as easily as if he were on *terra firma.* In the beginning, however, one experiences great difficulty in adjusting himself to the changes of perspective found in the air.

After this the pilot is sent back to Pau, where he has to perfect himself sufficiently in his art to master the various stunts essential in combat-work. Until then he may not go to the front for service in an "*appareil de chasse.*"

The first test is looping the loop. The machine is made to dive very fast for a short distance. Then the pilot gives a sharp pull on his controls, which makes it climb very abruptly, at the same time shutting off the motor. The little Nieuport climbs until it loses its speed, and then falls over backward. At the instant of reaching the line of diving the spark is then turned on again and the flight is resumed. The next requirement involves corkscrew looping, or, as they say in French, "*le renversement sur l'aile.*" This requires still greater skill than the previous test. It is not an easy manoeuvre to explain and, besides, I have not yet attempted it myself. The theory, however, is as follows: If you tip your machine enough to fly in a vertical position, your controls become reversed; in other words, the control for climbing and diving becomes the rudder and the rudder becomes the climbing control. To do the "*renversement*" the machine is put in a vertical position and the spark is shut off. The machine then loses its momentum and starts to fall. At that moment you must give a pull on your control and push the rudder "hard a-port," as a sailor would say. This forces the machine to complete the turn and dive from the normal horizontal position.

The final examination for the second brevet involves the dreaded "*vrille,*" or tail-spin. For many years any aviator who engaged in a *vrille* was given up for lost. Even today many aviators are killed attempting to master this most important trick, (as at time of first publication). Yet it has to be learned, for in modern aerial warfare it may sometime be the one manoeuvre which will enable you to escape from an assailant or make a sudden attack. The modern aeroplane is so stable that when it is made to dive it always attempts to rise and resume its flight. In the *"vrille,"* on the other hand, this resistance is overcome, and the machine spins down with incredible rapidity. The beginner usually commences by making one turn. He allows his machine to lose its speed and slip off on the wing. After engaging in a spiral, instead of continuing he then resumes his flight. The second time two turns have to be made.

More and more are made until the pilot feels that he has mastered the trick to his satisfaction. The first turn is usually made very slowly, but after that the speed increases with each succeeding turn until the machine is spinning on the comer of one wing as an axis. I have seen the more brilliant pilots at the front make as many as seven or eight turns, while they fell as far as five thousand feet. Every time I have seen anyone doing a *"vrille"* I have thought of the young lieutenant who was killed at Pau when I attended the school for the first time. What are dangers for the beginner, in the hands of the expert become weapons.

On completing this final course the pilot has learned everything that his instructors can teach him. It remains only for him to prove that in action he can avail himself of all the tricks that he has mastered. He has a machine that can manoeuvre to the best advantage, and he will enjoy a superiority which he never possessed with the heavier Farman biplane. Often I thought of this when flying over Verdun in the artillery machines. The little Nieuports seemed to circle about with such ease, doing whatever they pleased, while we lumbered about in constant danger of feeing attacked by some fast-flying *"Fritzie"* from the enemy's lines.

The principal task assigned to our *"avions de chasse"* is to keep the German airmen away from the French lines, and of attacking them when the opportunity offers. From an altitude of about thirteen thousand feet the Nieuports maintain a constant vigil. Although so small they are in fact the protectors of the larger artillery and reconnoissance machines. Far within the German lines several of the enemy's artillery biplanes are flying low. Farther up their fighting planes are waiting for an opportunity of coming over to attack the French. The shrapnel-puffs from our own guns reveal that someone is crossing our lines. A German artillery machine is coming to make a "*réglage*." One of the Fokkers is flying high above it, but the Nieuports are doing "ceiling work" and will look out for the intruders.

Different models of aeroplanes have a different position for their *mitrailleuses*. The attacking pilot always tries to find out from where he can make his attack without being riddled by his opponent. The proper position being obtained, the Nieuport is quickly turned toward its prey and at fifty yards the machine gun begins its *staccato* bark. To simplify the pilot's task the guns are always mounted in a fixed position and aimed dead ahead. Thus the pilot has only to think about pointing his own machine at the enemy. If he had to fly one way and shoot another he would be placed in a most disadvantageous position.

Combatants pass each other at terrific speed. There is time only for a few shots. If a hit is not scored during the first encounter, the attacking pilot goes through the same manoeuvre a second time. In the meanwhile the German airman is also doing his best to catch his opponent unawares. If the enemy succeeds in getting the Nieuport into a trap, then is the moment when he can put himself "*en vrille*" and escape.

Such is the course of training imposed upon every airman in France. It is the system which has been perfected under war conditions from the lessons learned during two years of the most desperate air conflicts.

LEONAUR

ALSO FROM LEONAUR
AVAILABLE IN SOFTCOVER OR HARDCOVER WITH DUST JACKET

THE 9TH—THE KING'S (LIVERPOOL REGIMENT) IN THE GREAT WAR 1914 - 1918 *by Enos H. G. Roberts*—Mersey to mud—war and Liverpool men.

THE GAMBARDIER *by Mark Severn*—The experiences of a battery of Heavy artillery on the Western Front during the First World War.

FROM MESSINES TO THIRD YPRES *by Thomas Floyd*—A personal account of the First World War on the Western front by a 2/5th Lancashire Fusilier.

THE IRISH GUARDS IN THE GREAT WAR - VOLUME 1 *by Rudyard Kipling*—Edited and Compiled from Their Diaries and Papers—The First Battalion.

THE IRISH GUARDS IN THE GREAT WAR - VOLUME 1 *by Rudyard Kipling*—Edited and Compiled from Their Diaries and Papers—The Second Battalion.

ARMOURED CARS IN EDEN *by K. Roosevelt*—An American President's son serving in Rolls Royce armoured cars with the British in Mesopatamia & with the American Artillery in France during the First World War.

CHASSEUR OF 1914 *by Marcel Dupont*—Experiences of the twilight of the French Light Cavalry by a young officer during the early battles of the great war in Europe.

TROOP HORSE & TRENCH *by R.A. Lloyd*—The experiences of a British Lifeguardsman of the household cavalry fighting on the western front during the First World War 1914-18.

THE EAST AFRICAN MOUNTED RIFLES *by C.J. Wilson*—Experiences of the campaign in the East African bush during the First World War.

THE LONG PATROL *by George Berrie*—A Novel of Light Horsemen from Gallipoli to the Palestine campaign of the First World War.

THE FIGHTING CAMELIERS *by Frank Reid*—The exploits of the Imperial Camel Corps in the desert and Palestine campaigns of the First World War.

STEEL CHARIOTS IN THE DESERT *by S. C. Rolls*—The first world war experiences of a Rolls Royce armoured car driver with the Duke of Westminster in Libya and in Arabia with T.E. Lawrence.

WITH THE IMPERIAL CAMEL CORPS IN THE GREAT WAR *by Geoffrey Inchbald*—The story of a serving officer with the British 2nd battalion against the Senussi and during the Palestine campaign.